초보자도 쉽게 이해할 수 있는

디지털회로
기초실험

초보자도 쉽게 이해할 수 있는

디지털회로
기초실험

박준식 지음

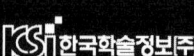

KSi 한국학술정보㈜

| 머리말

오늘날은 디지털기술의 발달에 따라 우리가 흔히 접하는 가전제품에서부터 컴퓨터, 제어용 기기, 의용 기기 등 산업용 기기에 이르기까지 거의 대부분의 전자시스템이 디지털화되어 있으며, 따라서 전기, 전자, 통신 등 전자공학의 관련 분야를 공부하는 학생들에게는 디지털공학에 대한 지식은 필수적이라 해도 과언이 아닐 것이다. 이 책은 디지털공학 기초회로 및 그 응용회로를 다루어 봄으로써 디지털공학에서 습득한 이론을 실험을 통해 이해할 수 있도록 하기 위해 체계적으로 구성해 놓았다.

이 책은 대학에서 한 학기 강의에 적합하도록 실험과제 단위로 구분하여 3부로 구성되어 있으며, 제1부 논리회로와 부울대수는 기본 논리 게이트 및 가감산기, 디코더, 멀티플렉서 및 플립플롭의 동작 특성을 실험토록 하였고, 제2부 카운터와 레지스터는 동기 및 비동기 카운터, Up/Down 카운터, 링 카운터 및 시프트레지스터로 구성되어 카운터의 동작 특성을 완벽히 이해할 수 있도록 하였으며, 제3부 응용회로 및 제작 부분에서는 D/A, A/D 변환기 및 LED와 카운터를 이용한 스톱워치로 구성하여 실제 시스템을 제작해 봄으로써 디지털 시스템을 이해하는 데 도움이 되도록 하였다. 각 실험별로 항상 기초적인 개념을 설명하고 실험에 임할 수 있도록 하였기 때문에 처음 접하는 학생도 실험에 별 어려움이 없을 것으로 생각된다.

나름대로는 충실히 한다고 했으나 부족한 점이 많을 것으로 생각되지만 아무쪼록 본서를 통해 디지털공학을 이해하는 데 조금이라도 도움이 되었으면 하는 바람을 가지고 있다. 독자 여러분의 기탄없는 충고를 바라며, 본서의 출판을 위해 수고해 주신 출판사 사장님을 비롯한 임직원 여러분께 감사드린다. 끝으로 본서 집필을 위해 연구실에서 방학과 휴일을 반납한 본인에게 불평 한 마디 하지 않고 격려해 준 아내에게 진심으로 감사의 말을 전하고 싶다.

당신 곁에는 희생 플라이 하나쯤 날려줄 그녀가 있는가?

- 금산골 연구실에서 저자 씀-

차례

품목별 소요재료 총괄표

품명	규격	실험1	실험2	실험3	실험4	실험5	실험6	실험7	실험8	실험9	실험10	실험11	실험12	실험13	실험14	실험15	계
가변저항	10kΩ													1			1
가변저항	500kΩ														1	1	2
저항	100kΩ														6		6
저항	10kΩ													2			2
저항	1kΩ						2	3									5
저항	200kΩ														5		5
저항	20kΩ													1			1
저항	300Ω															21	21
저항	30kΩ													1			1
저항	32kΩ															1	1
저항	330Ω			2	7			4									13
저항	3kΩ	1	1														2
저항	40kΩ													1			1
저항	680Ω						1								1		2
저항	8.2kΩ															1	1
저항	80kΩ													1			1
저항	820Ω															2	2
컨덴서	0.012μF															1	1
컨덴서	0.033μF															1	1
컨덴서	0.03μF						1										1
컨덴서	22μF															1	1
AND	74LS08	1	1		1					1	1						5
AND	74LS11		1							1	1						3
AND	74LS15	1															1
Counter	74LS193										1						1
Counter	74LS90															4	4
Counter	74LS93													1			1
Counter	MC14510														1		1
Decoder	74LS47				1											3	4
EX - NOR	74LS266	1															1
EX - OR	74LS86	1		1													2
FND 500	7 - Segment				1											3	4
JK FF	74LS73								2	2	2						6
JK FF	74LS76							1				2	2			1	6
LED				2				4							1		7
M/V	74LS121															1	1
NAND	74LS00	1	1	1			2	3	1			1	2			1	13
NAND	74LS10		1			2		2				1					6
NAND	74LS20					3											3
NAND	74LS38	1															1
NAND	MC14011														1		1
NOR	74LS02	1	1	1			1	1									5
NOT	74LS04	1	1	1		1		1					1				6
NOT	74LS05	1	1														2
OP AMP	CA3160														1		1
OP AMP	μ741													1			1
OR	74LS32	1	1		1						1						4
PB Sw																2	2
슈미트트리거	74LS14															1	1
Timer	555															1	1

실험별 소요재료 총괄표

품명	규격	실험1	실험2	실험3	실험4	실험5	실험6	실험7	실험8	실험9	실험10	실험11	실험12	실험13	실험14	실험15
NOT	74LS04	1	1	1		1		1					1			
NOR	74LS02	1	1	1			1	1								
NAND	74LS00	1	1	1			2	3	1			1	2			1
AND	74LS08	1	1		1					1	1					
OR	74LS32	1	1		1						1					
저항	3kΩ	1	1													
NOT	74LS05	1	1													
EX－OR	74LS86	1		1												
AND	74LS15	1														
EX－NOR	74LS266	1														
NAND	74LS38	1														
NAND	74LS10		1			2		2				1				
AND	74LS11		1							1	1					
저항	330Ω			2	7			4								
LED				2				4							1	
Decoder	74LS47				1											3
FND 500	7－Segment				1											3
NAND	74LS20					3										
저항	680Ω						1								1	
컨덴서	0.03μF						1									
저항	1kΩ						2	3								
JK FF	74LS76							1				2	2			1
JK FF	74LS73								2	2	2					
Counter	74LS193										1					
가변저항	10kΩ													1		
저항	20kΩ													1		
저항	30kΩ													1		
저항	40kΩ													1		
저항	80kΩ													1		
Counter	74LS93													1		
OP AMP	μ741													1		
저항	10kΩ													2		
가변저항	500kΩ														1	1
Counter	MC14510														1	
NAND	MC14011														1	
OP AMP	CA3160														1	
저항	200kΩ														5	
저항	100kΩ														6	
저항	32kΩ															1
저항	8.2kΩ															1
컨덴서	0.012μF															1
컨덴서	0.033μF															1
컨덴서	22μF															1
M/V	74LS121															1
슈미트트리거	74LS14															1
Timer	555															1
저항	820Ω															2
PB SW																2
Counter	74LS90															4
저항	300Ω															21

인생의 **홈런**을 기다리는가? 그렇다면 **오늘**부터 꾸준히 **안타**를 **쳐** 나가라!

제1부   논리회로와 부울대수

함께하는 사람이 소중하다는 것, 레드카드로 나가는 동료의 등 뒤에서 읽어라!

실험 1 기본 논리 게이트

실험목적

- AND, OR, NOT, NAND, NOR, EX－OR, EX－NOR 게이트의 기본적인 동작원리 및 논리함수를 이해하도록 한다.
- 논리소자의 효율적인 사용을 위해 TTL 게이트의 특성을 이해하도록 한다.
- 논리회로의 측정방법을 이해하도록 한다.

실험기기 및 재료

구분	품명	규격	수량	비고
기기	논리회로 실험장치		1	
	회로시험기		1	
재료	AND 게이트	IC 74LS08	1	
	AND 게이트	IC 74LS15	1	open collector
	OR 게이트	IC 74LS32	1	
	NOT 게이트	IC 74LS04	1	
	NOT 게이트	IC 74LS05	1	open collector
	NAND 게이트	IC 74LS00	1	
	NAND 게이트	IC 74LS38	1	open collector
	NOR 게이트	IC 74LS02	1	
	EX－OR 게이트	IC 74LS86	1	
	EX－NOR 게이트	IC 74LS266	1	open collector
	저항	3kΩ	1	
	점퍼선		약간	

1. AND 게이트

AND 게이트는 모든 입력이 High일 경우에만 출력이 High가 되는 게이트를 말한다. AND 게이트는 그림 1-1과 같은 기호로 표시하며 진리표는 표 1-1과 같다. 따라서 2입력 AND 게이트의 논리식은 $F = A \cdot B$로 나타낼 수 있다.

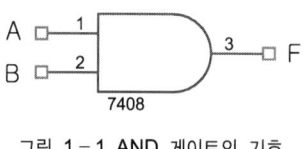

그림 1-1 AND 게이트의 기호

표 1-1 AND 게이트의 진리표

A	B	F
0	0	0
0	1	0
1	0	0
1	1	1

TTL AND 게이트는 2개의 입력 단자를 가진 것으로 4개의 게이트가 1개의 IC 안에 수용된 7408형(quad 2-input AND gate)이 있으며 3입력 AND 게이트는 7411, 4입력 AND 게이트는 7421이 있다. C-MOS IC는 4081(2-input), 4073(3-input), 4082(4-input)가 있다. 그림 1-2와 같이 AND 게이트를 접속하여 입력 단자 수를 증가시킬 수가 있다.

그림 1-3은 다이오드를 이용한 AND 게이트를 나타내고 있다. 이 그림에서 A=B=0, A=0과 B=1, A=1과 B=0인 경우는 다이오드가 모두 ON 상태로 되어 출력은 0이 되고 A=B=1인 경우에만 모든 다이오드가 OFF 되어 출력이 1이 되고 AND 회로로 동작한다.

그림 1-4는 AND 게이트 IC 칩의 예이다. 이것은 표준 TTL 계열에서 SN7408을 의미하며, 1개의 칩 내에 AND 게이트 4개가 들어 있다.

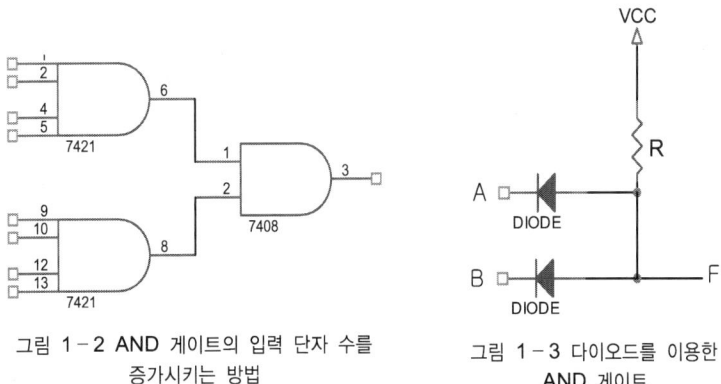

그림 1-2 AND 게이트의 입력 단자 수를 증가시키는 방법

그림 1-3 다이오드를 이용한 AND 게이트

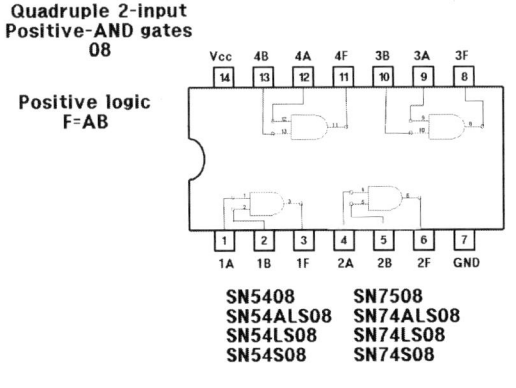

그림 1-4 AND 게이트 IC

2. OR 게이트

OR 게이트는 입력이 하나라도 High일 경우에는 출력이 High가 되는 게이트를 말하며, OR 게이트의 기호와 진리표는 그림 1-5, 표 1-2와 같다. 따라서 논리식은 F＝A＋B로 나타낼 수 있다. TTL IC로 OR 게이트는 7432(quad 2-input OR)형이 있으며, C-MOS IC로는 4071(quad 2-input OR), 4072(quad 4-input OR)형이 있다. OR 게이트의 입력 단자 수를 늘리기 위해서는 그림 1-2의 AND 게이트의 경우와 같이 하면 된다.

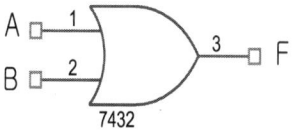

그림 1-5 OR 게이트의 기호

표 1-2 OR 게이트의 진리표

A	B	F
0	0	0
0	1	1
1	0	1
1	1	1

그림 1-6은 다이오드를 이용한 OR 게이트를 나타내고 있다. 이 그림에서 A＝B＝0이면 다이오드에 걸리는 전압이 없으므로 출력 F는 0이 되고, A＝0과 B＝1 또는 A＝1과 B＝0 또는 A＝B＝1인 경우에는 다이오드의 어느 하나나 모두에 전압이 걸려 출력은 1이 나와 OR 게이트로서 동작한다.

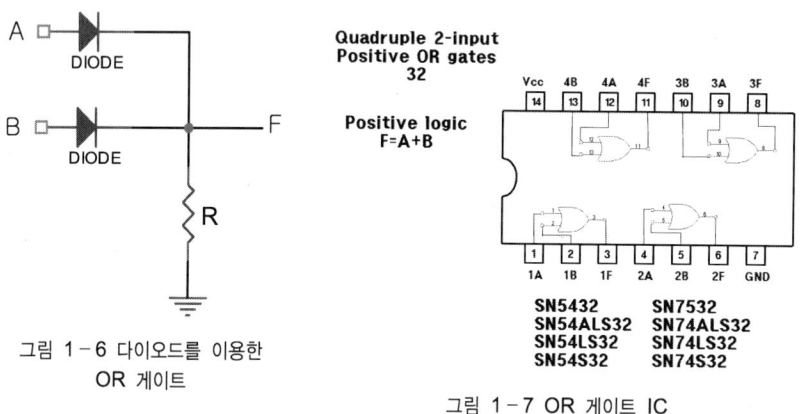

그림 1-6 다이오드를 이용한 OR 게이트

그림 1-7 OR 게이트 IC

3. NOT 게이트

NOT 게이트는 부정회로 또는 Inverter라고도 하며 입력값을 반전시키는 역할을 한다. NOT 게이트의 논리기호와 진리표는 그림 1-8, 표 1-3과 같다. 이 게이트는 하나의 입력과 출력을 가지며 입력이 High이면 출력은 Low, 입력이 Low이면 출력은 High가 되어 논리식은 $F = \overline{A}$로 나타낼 수 있다. 이 게이트는 주로 다른 게이트의 입, 출력에 연결하며 어떤 상태를 반전시키는 역할을 한다.

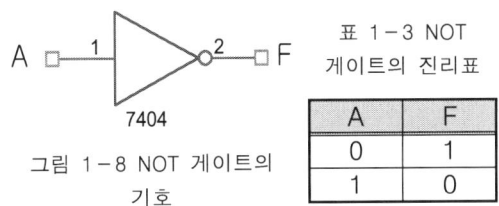

표 1-3 NOT 게이트의 진리표

A	F
0	1
1	0

그림 1-8 NOT 게이트의 기호

그림 1-9는 트랜지스터를 이용한 NOT 회로를 나타내며 입력 A=1이면 트랜지스터가 ON 되어 출력 F=0이 되고, A=0이면 트랜지스터가 OFF 되어 출력 F=1이 된다. 게이트를 종속 접속하면 게이트를 통과하는 신호의 전송 시간을 증가시키게 되므로 NOT 게이트를 여러 개 종속 접속하여 지연회로나 상태 반전회로로 사용하기도 하며 총 지연 시간은 게이트 하나의 전송 시간에 전 게이트의 수를 곱해 주면 된다. TTL IC로는 7404형이 있으며 6개의 Inverter가 IC 안에 들어가 있다. 7405형은 Open collector형의 Inverter이다. C-MOS IC로는 4009형이 있다.

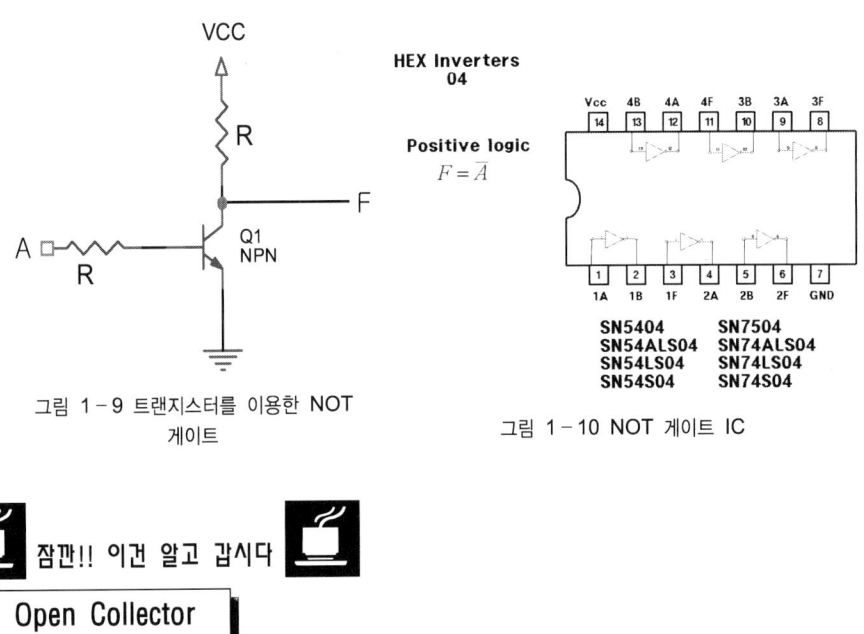

그림 1-9 트랜지스터를 이용한 NOT 게이트

그림 1-10 NOT 게이트 IC

![잠깐!! 이건 알고 갑시다]

Open Collector

Wired-AND를 구현하기 위하여, 어떤 TTL 회로는 collector 측을 개방한 형태의 개방-컬렉터(open-collector) 출력을 갖도록 설계되는데, 아래 그림에서 출력이

Low 상태에서 Q4는 ON(베이스 전류를 가짐) 되고, 출력이 High 상태에서 Q4는 Off 된다. 이 회로를 적절하게 동작시키기 위해서는 collector 측에 출력 저항을 연결해야 하는데 이 저항을 풀-업(pull-up) 저항이라고 한다. 이 IC는 출력 단에서 전원전압이 다른 논리회로와 연결 시 주로 사용된다.

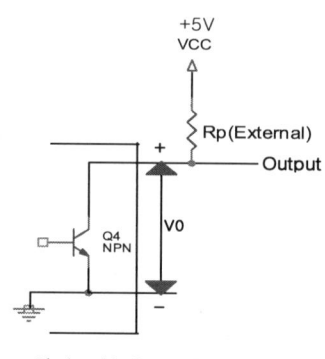

그림 1-11 Open Collector 회로

4. NAND 게이트

NAND 게이트는 디지털 게이트 중에서 가장 많이 사용되는 게이트이며 AND 게이트에 Inverter에 접속시킨 것과 같은 것으로 모든 입력이 High일 경우에만 출력이 Low가 되는 논리 게이트로서 논리식은 $F = \overline{A \cdot B}$와 같이 표현할 수 있다. 그림 1-12와 표 1-4는 NAND 게이트의 논리기호와 진리표를 나타낸다.

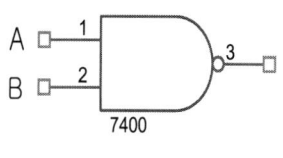

그림 1-12 NAND 게이트의 기호

표 1-4 NAND 게이트의 진리표

A	B	F
0	0	1
0	1	1
1	0	1
1	1	0

대표적인 NAND 게이트는 7400(quad 2-input NAND), 7401(quad 2-input NAND, open collector), 7410(triple 3-input NAND), 7420(dual 4-input NAND), 7430(8-input NAND) 등이 있고, C-MOS IC로는 4011(quad 2-input NAND), 4023(triple 3-input NAND), 4012(dual 4-input NAND), 4086(8-input NAND) 등이 있다.

그림 1-13은 DTL 기본 NAND 게이트를 나타내며, 이 회로에서 입력 A, B, C가 모두 High 상태이면 트랜지스터 Q는 포화 영역으로 구동되어 출력 F는 0이 된

다. E점의 전압은 VBE와 D1, D2를 가로지르는 2개의 다이오드 전압 강하와의 합과 같다(0.7×3 = 2.1[V]) 모든 입력이 5[V]이고 E점의 전압이 2.1[V]이므로 입력단의 다이오드는 차단된다. 이때 베이스 전류 IB는 흐르게 될 것이고 트랜지스터는 포화되어 출력 F는 VCE(sat) = 0.2[V], 즉 Low 상태가 된다. 3입력 중 어느 하나라도 Low 상태이면 E점의 전압은 입력전압 0.2[V]에 다이오드 전압 강하 0.7[V]를 합한 0.9[V]가 된다. 트랜지스터가 도통하기 위해서는 E점의 전압이 VBE 전압 강하와 2개의 다이오드 D1, D2의 전압 강하의 합 0.6×3 = 1.8[V]를 넘어야 한다. 그런데 E점의 전압은 0.9[V]이므로 트랜지스터 Q는 차단되고 출력은 High가 된다.

그림 1-14에 나타낸 것은 TTL 시리즈의 SN7400으로 2입력 NAND 게이트가 4개 들어 있다.

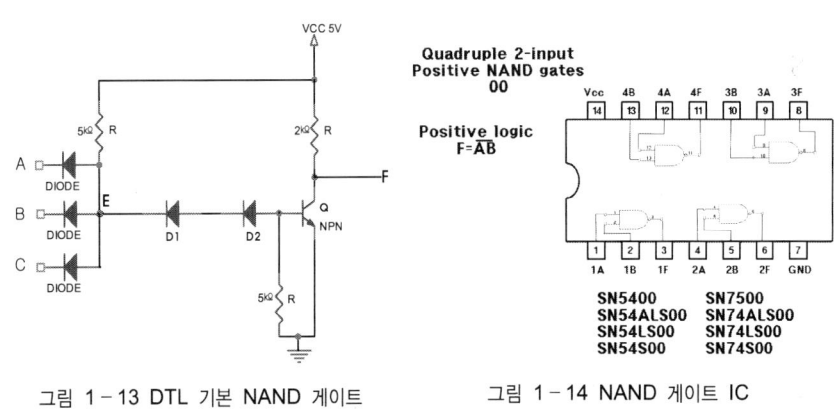

그림 1-13 DTL 기본 NAND 게이트   그림 1-14 NAND 게이트 IC

5. NOR 게이트

NOR 게이트는 NAND 게이트와 함께 많이 사용되는 게이트로 OR 게이트의 출력을 NOT 게이트로 반전시키는 회로이다. 이 게이트는 OR 게이트와 반대로 어느 입력 하나라도 High이면 출력이 Low가 되는 게이트이다. 논리식은 $F = \overline{A+B}$와 같이 표현할 수 있다. 그림 1-15와 표 1-5는 NOR 게이트의 논리기호와 진리표이다.

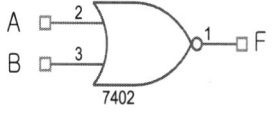

그림 1-15 NOR 게이트의
기호

표 1-5 NOR게이트의 진리표

A	B	F
0	0	1
0	1	0
1	0	0
1	1	0

NOR 게이트의 TTL IC로는 7402(quad 2-input NOR), 7427(triple 3-input NOR), 7425(dual 4-input NOR) 등이 있고, C-MOS IC로는 4001(quad 2-input NOR), 4025(triple 3-input NOR), 4002(dual 4-input NOR) 등이 있다.

그림 1-16은 대표적인 NOR 게이트 IC를 보여준다.

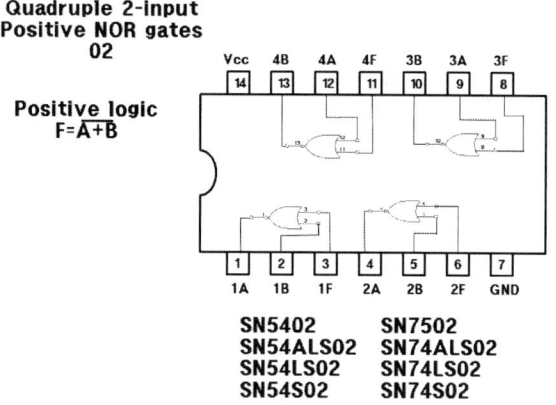

그림 1-16 NOR 게이트 IC

6. EX-OR 게이트

EX-OR(Exclusive OR) 게이트는 배타적 OR 게이트 또는 반일치 논리회로라고도 하며 High의 수가 짝수 개일 때 출력이 Low가 되며, 홀수 개일 때만 High가되는 회로를 말한다. EX-OR 게이트의 기호는 그림 1-17과 같이 표시하며 진리표는 표 1-6과 같다.

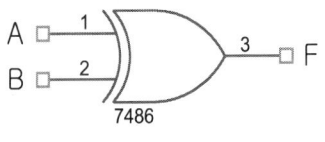

표 1-6 EX-OR게이트의 진리표

A	B	F
0	0	0
0	1	1
1	0	1
1	1	0

그림 1-17 EX-OR 게이트의
기호

따라서 EX-OR의 논리식은

$F = \overline{A}B + A\overline{B} = A \oplus B$로 표시된다.

일반적으로 EX-OR 게이트는 기본 논리 게이트로 취급하지 않는 경우도 있으나 가·감산기, 비교기, 패리티 검출기 등 그 응용범위는 실로 넓다. TTL IC의 경우 7486(quad 2-input Exclusive OR)이 있으며 C-MOS IC로는 4030(quad 2-input Exclusive OR)이 있다.

EX-OR를 사용하면 2진 비트의 1의 총수가 짝수인가 홀수인가를 판별할 수 있는데 이 기능을 이용하여 검사기(Parity Checker)로 사용할 수가 있다. 즉 두 입력 중 1의 개수가 짝수 개이면 출력이 '0'이고 홀수 개이면 출력이 '1'이므로 2진 비트의 수가 홀수 개인지 짝수 개인지를 판별할 수가 있다.

그림 1-18은 4개의 비트 A, B, C, D의 패리티를 검사할 수 있는 EX-OR 게이트를 사용한 패리티 검사기이며, 그림 1-19와 1-20은 각각 짝수 패리티 검사기, 홀수 검사기 회로를 나타낸다.

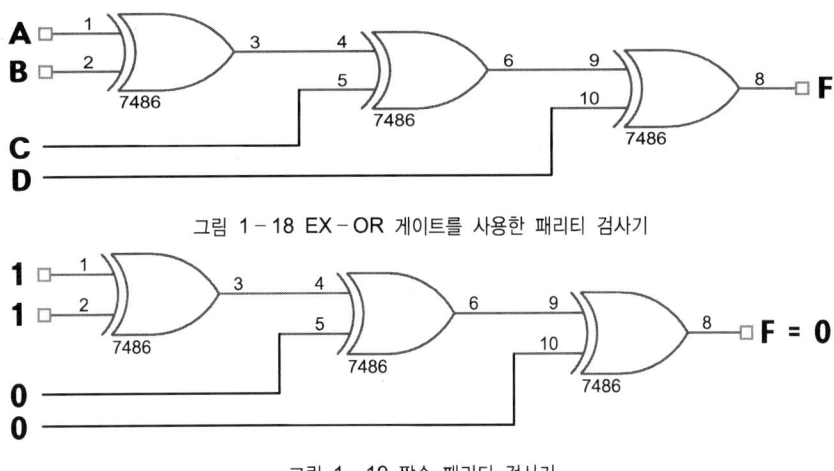

그림 1-18 EX-OR 게이트를 사용한 패리티 검사기

그림 1-19 짝수 패리티 검사기

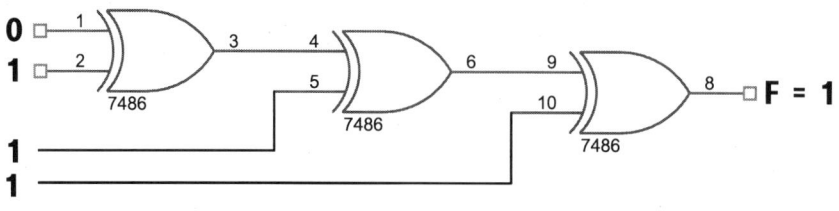

그림 1-20 홀수 패리티 검사기

4비트를 가산하는 마지막 단계인 출력 F가 0이면 짝수 패리티를 나타내고 1이면 홀수 패리티를 나타낸다. 그림 1-19는 ABCD=1100인 경우로 출력 F가 0이므로 짝수 패리티를, 그림 1-20은 ABCD=0111의 경우로 출력 F가 1이므로 홀수 패리티임을 나타낸다.

7. EX-OR 게이트

EX-NOR(Exclusive NOR) 게이트는 EX-OR 게이트를 부정하면 되며 배타적 NOR 게이트라고 한다. 이 게이트는 EX-OR 게이트와 반대로 1의 총수가 짝수이면 출력이 1이고 홀수이면 0이므로 논리등가 또는 일치회로라고도 한다. 이 게이트의 기호와 진리표는 아래와 같고, 논리식은

$$F = \overline{A \oplus B} = \overline{\overline{A}B + A\overline{B}} = \overline{A}\,\overline{B} + AB = A \odot B$$ 가 된다.

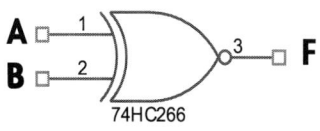

그림 1-21 EX-NOR 게이트의 기호

표 1-7 EX-NOR 게이트의 진리표

A	B	F
0	0	1
0	1	0
1	0	0
1	1	1

EX-NOR 게이트의 TTL IC로는 74266(quad 2-input EX-NOR with OC), 74260(dual 5-input EX-NOR)가 있고, C-MOS IC로는 4077(quad 2-input EX-NOR)가 있다.

본 교재에서 사용하는 대부분의 IC는 아래 그림과 같은 Dual in-line형 타입을 사용한다. Dual in-line형 IC의 핀 번호는 IC를 위에서 내려 보았을 때 그림에서 보는 바와 같다. 핀 1번을 표시하기 위해서 점(Dot)이 있는 것도 있고 한쪽이 파인 것도 있다.

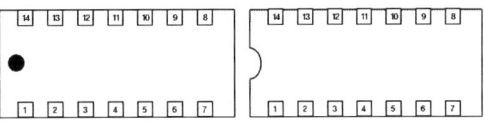

1. AND 게이트

(1) 그림 1-22와 같이 회로를 구성하고 출력전압을 측정하여 표 1-8에 기록하시오.

표 1-8

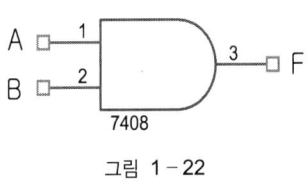

그림 1-22

A	B	F
0	0	
0	1	
1	0	
1	1	

(2) 그림 1 – 23과 같이 회로를 구성하고 출력전압을 측정하여 표 1 – 9에 기록하고 2입력 AND 게이트의 출력과 비교하시오.

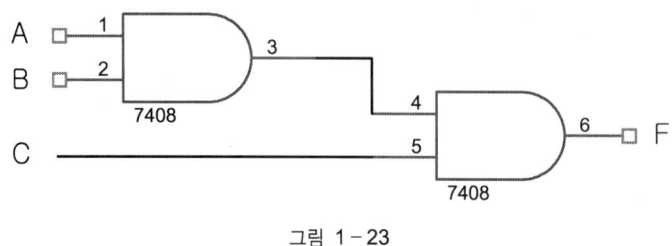

그림 1 – 23

표 1 – 9

A	B	C	F
0	0	0	
0	0	1	
0	1	0	
0	1	1	
1	0	0	
1	0	1	
1	1	0	
1	1	1	

(3) 그림 1-24와 같이 회로를 구성하고 출력전압을 측정하여 표 1-10에 기록하
시오.

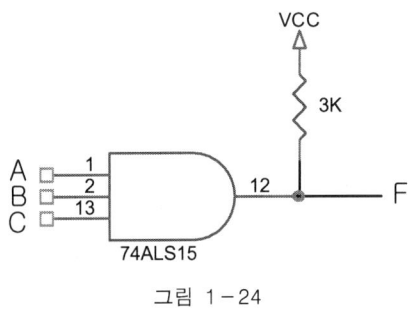

그림 1-24

표 1-10

A	B	C	F
0	0	0	
0	0	1	
0	1	0	
0	1	1	
1	0	0	
1	0	1	
1	1	0	
1	1	1	

2. OR 게이트

(4) 그림 1-25와 같이 회로를 구성하고 출력전압을 측정하여 표 1-11에 기록하시오.

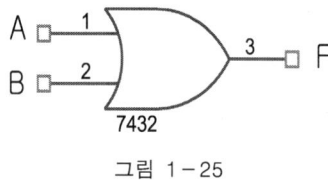

그림 1-25

표 1-11

A	B	F
0	0	
0	1	
1	0	
1	1	

(5) 그림 1 – 26과 같이 회로를 구성하고 출력전압을 측정하여 표 1 – 12에 기록하고 2입력 OR 게이트와 비교하여 보시오.

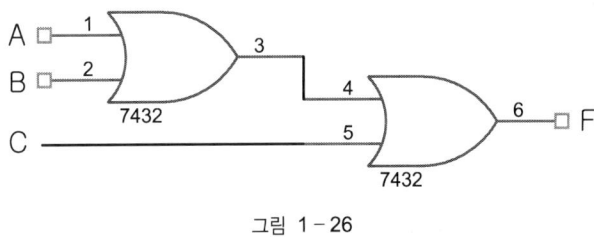

그림 1 – 26

표 1 – 12

A	B	C	F
0	0	0	
0	0	1	
0	1	0	
0	1	1	
1	0	0	
1	0	1	
1	1	0	
1	1	1	

3. NOT 게이트

(6) 그림 1-27과 같이 회로를 구성하고 출력전압을 측정하여 표 1-13에 기록하시오.

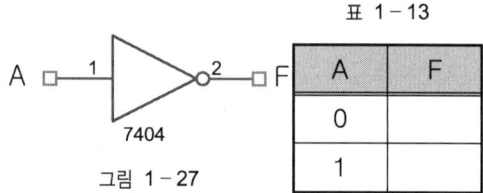

표 1-13

A	F
0	
1	

그림 1-27

(7) 그림 1-28과 같이 회로를 구성하고 출력전압을 측정하여 표 1-14에 기록하시오.

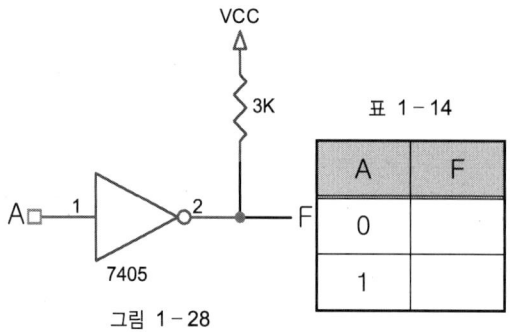

표 1-14

A	F
0	
1	

그림 1-28

4. NAND 게이트

(8) 그림 1−29와 같이 회로를 구성하고 출력전압을 측정하여 표 1−15에 기록하시오.

표 1−15

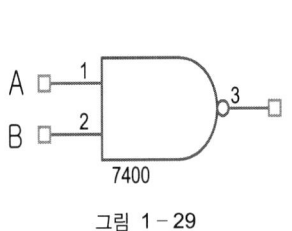

그림 1−29

A	B	F
0	0	
0	1	
1	0	
1	1	

(9) 그림 1−30과 같이 회로를 구성하고 출력전압을 측정하여 표 1−16에 기록하시오.

표 1−16

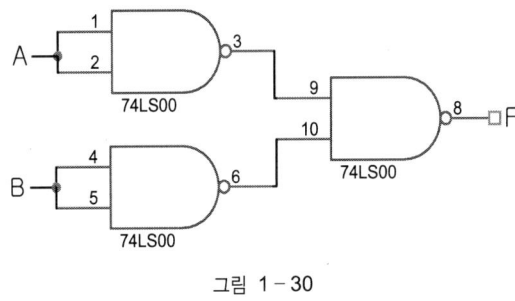

그림 1−30

A	B	F
0	0	
0	1	
1	0	
1	1	

(10) 그림 1 - 31과 같이 회로를 구성하고 출력전압을 측정하여 표 1 - 17에 기록
하고 2입력 NAND 게이트와 비교하여 보시오.

표 1 - 17

A	B	C	F
0	0	0	
0	0	1	
0	1	0	
0	1	1	
1	0	0	
1	0	1	
1	1	0	
1	1	1	

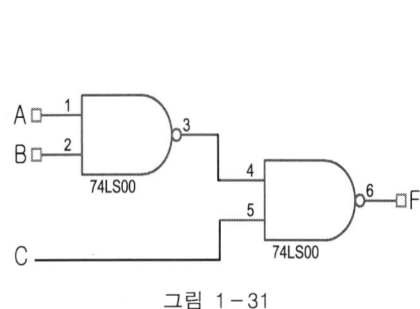

그림 1 - 31

(11) 그림 1 - 32와 같이 회로를 구성하고 출력전압을 측정하여 표 1 - 18에 기록
하시오.

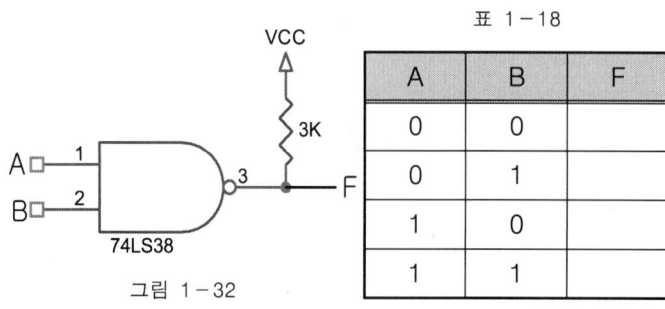

그림 1 - 32

표 1 - 18

A	B	F
0	0	
0	1	
1	0	
1	1	

5. NOR 게이트

(12) 그림 1−33과 같이 회로를 구성하고 출력전압을 측정하여 표 1−19에 기록하시오.

표 1−19

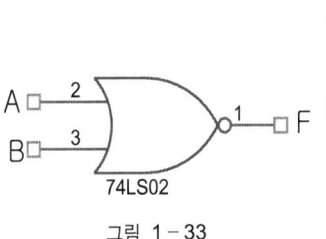

그림 1−33

A	B	F
0	0	
0	1	
1	0	
1	1	

(13) 그림 1−34와 같이 회로를 구성하고 출력전압을 측정하여 표 1−20에 기록하시오.

표 1−20

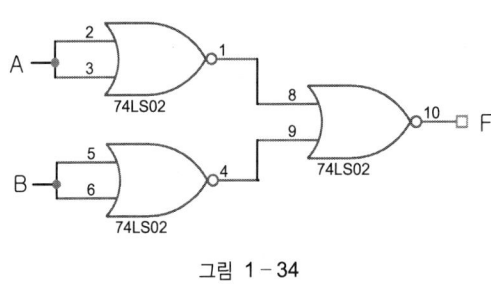

그림 1−34

A	B	F
0	0	
0	1	
1	0	
1	1	

6. E-OR 및 E-NOR 게이트

(14) 그림 1-35와 같이 회로를 구성하고 출력전압을 측정하여 표 1-21에 기록하시오.

표 1-21

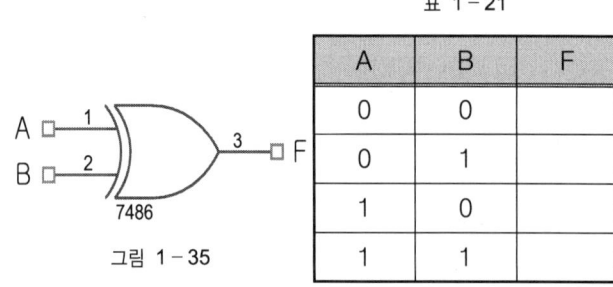

그림 1-35

A	B	F
0	0	
0	1	
1	0	
1	1	

(15) 그림 1-36과 같이 회로를 구성하고 출력전압을 측정하여 표 1-22에 기록하시오.

표 1-22

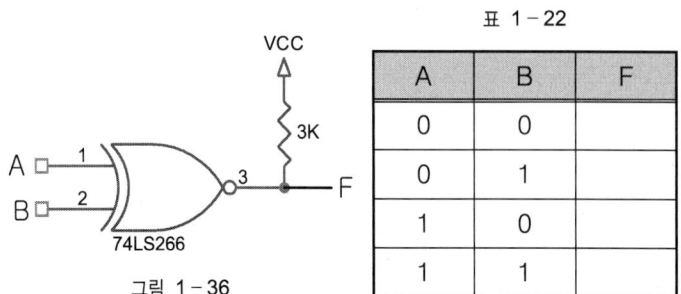

그림 1-36

A	B	F
0	0	
0	1	
1	0	
1	1	

실험결과 Report	학과명	학번	성명
실험 1 기본 논리 게이트			

1. AND 게이트

표 1-9

A	B	C	F
0	0	0	
0	0	1	
0	1	0	
0	1	1	
1	0	0	
1	0	1	
1	1	0	
1	1	1	

표 1-8

A	B	F
0	0	
0	1	
1	0	
1	1	

표 1-10

A	B	C	F
0	0	0	
0	0	1	
0	1	0	
0	1	1	
1	0	0	
1	0	1	
1	1	0	
1	1	1	

2. OR 게이트

표 1-12

A	B	C	F
0	0	0	
0	0	1	
0	1	0	
0	1	1	
1	0	0	
1	0	1	
1	1	0	
1	1	1	

표 1-11

A	B	F
0	0	
0	1	
1	0	
1	1	

3. NOT 게이트

표 1-13

A	F
0	
1	

표 1-14

A	F
0	
1	

4. NAND 게이트

표 1-15

A	B	F
0	0	
0	1	
1	0	
1	1	

표 1-16

A	B	F
0	0	
0	1	
1	0	
1	1	

표 1-17

A	B	C	F
0	0	0	
0	0	1	
0	1	0	
0	1	1	
1	0	0	
1	0	1	
1	1	0	
1	1	1	

표 1-18

A	B	F
0	0	
0	1	
1	0	
1	1	

5. NOR 게이트

표 1-19

A	B	F
0	0	
0	1	
1	0	
1	1	

표 1-20

A	B	F
0	0	
0	1	
1	0	
1	1	

6. EX-OR 및 EX-NOR 게이트

표 1-21

A	B	F
0	0	
0	1	
1	0	
1	1	

표 1-22

A	B	F
0	0	
0	1	
1	0	
1	1	

드리블이 잘 나갈 때 **어디선가** 들어오는 **태클**을 **대비**하라!

실험 2	불 대수와 논리회로의 간소화

실험목적

- 불 대수의 공리와 정리를 이해하고 실험적으로 증명한다.
- 불 대수식을 논리회로로 나타낼 수 있는 능력을 키운다.
- 불 대수식을 이용한 논리회로의 간소화 방법을 이해하도록 한다.
- DE·morgan의 정리를 이해하고 활용할 수 있도록 한다.

실험기기 및 재료

구분	품명	규격	수량	비고
기기	논리회로 실험장치		1	
	회로시험기		1	
재료	AND 게이트	IC 74LS08	1	
	AND 게이트	IC 74LS11	1	
	OR 게이트	IC 74LS32	1	
	NOT 게이트	IC 74LS04	1	
	NOT 게이트	IC 74LS05	1	open collector
	NAND 게이트	IC 74LS00	1	
	NAND 게이트	IC 74LS10	1	
	NOR 게이트	IC 74LS02	1	
	저항	3kΩ	1	
	점퍼선		약간	

1. 불 대수(Boolean Algebra)

불 대수는 '0'과 '1' 2개의 요소와 · (AND), ＋(OR), －(NOT)의 3개의 연산자를 사용하는 대수로 다음과 같은 공리(公理)를 바탕으로 전개된다.

공리 1.

(1) $A \neq 0$이면 $A = 1$이고, $A = 1$이면 $\overline{A} = 0$

(2) $A \neq 1$이면 $A = 0$이고, $A = 0$이면 $\overline{A} = 1$

공리 2.

(1) $0 + 0 = 0$

(2) $0 + 1 = 1 + 0 = 1$

(3) $1 + 1 = 1$

공리 3.

(1) $0 \cdot 0 = 0$

(2) $0 \cdot 1 = 1 \cdot 0 = 0$

(3) $1 \cdot 1 = 1$

공리 4.

(1) $\overline{1} = 0$

(2) $\overline{0} = 1$

위의 불 대수를 증명할 수 있는 방법은 여러 가지가 있으나 참과 거짓, 있다(實)와 없다(虛), 스위치의 개폐, 전압의 고저 등으로 대응하여 설명할 수 있다. 위의 공리를 기초로 불 대수를 다음과 같이 정리할 수 있다.

정리 1. 교환법칙

(1) $A + B = B + A$

(2) $A \cdot B = B \cdot A$

정리 2. 결합법칙

(1) $(A \ B) \ C = A \ (B \ C)$

(2) $(A + B) + C = A + (B + C)$

정리 3. 분배법칙

(1) $(A + B)(A + C) = AA + AC + AB + BC = A(1 + B + C) + BC = A + BC$

(2) $AB + AC = A(B + C)$

정리 4. 동일법칙

(1) $A + A = A$

(2) $A \cdot A = A$

정리 5. 흡수법칙

(1) $A + (A \cdot B) = A(1 + B) = A$

(2) $A \cdot (A + B) = AA + AB = A + AB = A(1 + B) = A$

정리 6.

(1) $A \cdot 0 = 0$

(2) $A + 0 = A$

정리 7.

(1) $A \cdot 1 = A$

(2) $A + 1 = A$

정리 8.

(1) $A \cdot \overline{A} = 0$

(2) $A + \overline{A} = 1$

정리 9.

(1) $\overline{\overline{A}} = A$

2. 드모르간의 정리(De · Morgan's theorem)

NOR 게이트와 NAND 게이트는 OR나 AND 게이트에 NOT 게이트를 사용하면 등가회로를 얻을 수가 있는데 이 방법을 De · Morgan의 정리라고 한다. 이 정리는 NOR와 NAND 게이트를 이용한 응용 및 논리회로를 간소화시키는 데 널리 이용할 수 있다. 실제 IC 논리회로는 NOR 및 NAND 게이트만을 이용해서 회로를 구성하는 경우가 많이 있는데 이는 동일 회로의 조합에 따라서 모든 논리를 얻을 수 있어 회로를 단순화하기가 쉽고 대량 생산에 적합하기 때문이다. 이 정리를 확실히 이해하여 회로를 단순화하는 데 도움이 되기 바란다.

De · Morgan의 정리를 요약하면 아래와 같다.

(1) $\overline{A + B} = \overline{A} \cdot \overline{B}$

논리합의 부정은 각 변수들의 부정의 논리곱과 같다.

(2) $\overline{A \cdot B} = \overline{A} + \overline{B}$

논리곱의 부정은 각 변수들의 부정의 논리합과 같다.

위의 De · Morgan의 정리를 논리 게이트로 표시하면 아래 그림과 같다.

(1) De·Morgan의 제1정리

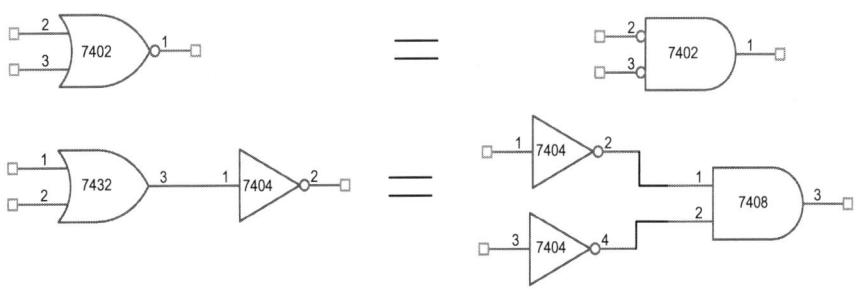

그림 2-1 De · Morgan의 제1정리

(2) De·Morgan의 제2정리

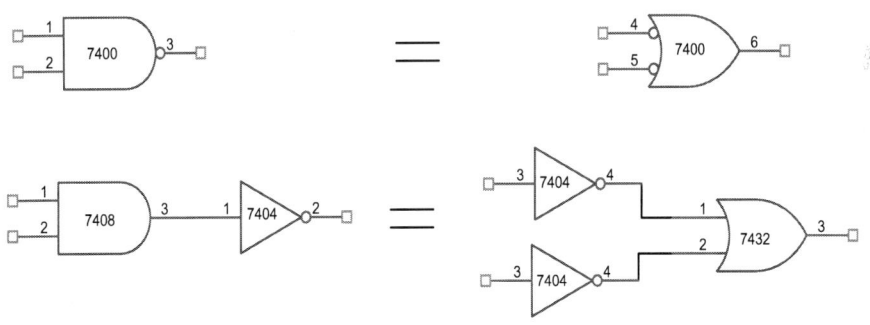

그림 2-2 De · Morgan의 제2정리

3. 논리회로의 간소화

  De · Morgan의 정리를 이용하면 NOR 게이트나 NAND 게이트만으로 기본 논리 회로를 간소화할 수 있는데 이는 NOR나 NAND 게이트를 가지고 기본 논리 게이트, 즉 AND, OR, NOT 회로를 실현할 수 있기 때문이다. 이제 NOR와 NAND 게이트가 어떻게 AND, OR, NOT으로 사용될 수 있는가 살펴보기로 하자.

(1) NOR 게이트만을 이용한 기본 논리회로

$$A + B = \overline{\overline{A + B}}$$

$$A \cdot B = \overline{\overline{A \cdot B}} = \overline{\overline{A} + \overline{B}}$$

이므로 NOR 게이트만을 이용하여 NOT, OR, AND, NAND 게이트를 구성하면 다음 그림과 같다.

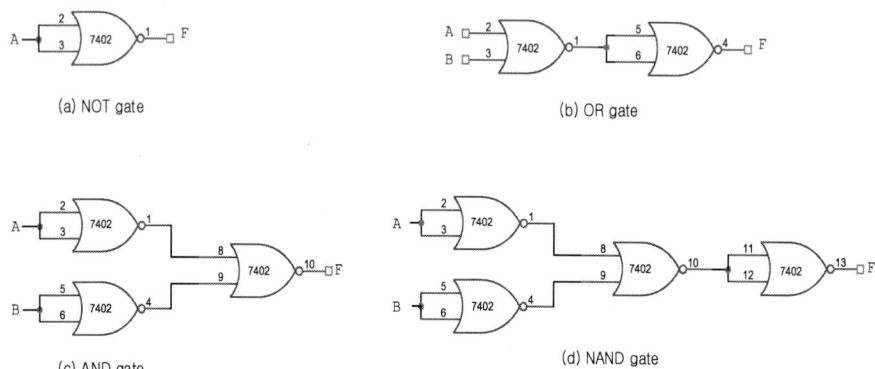

(a) NOT gate   (b) OR gate

(c) AND gate   (d) NAND gate

그림 2-3 NOR 게이트만을 이용한 기본 논리회로

(2) NAND 게이트만을 이용한 기본 논리회로

$$A \cdot B = \overline{\overline{A \cdot B}}$$

$$A + B = \overline{\overline{A + B}} = \overline{\overline{A} \cdot \overline{B}}$$

이므로 NAND 게이트만을 이용하여 NOT, OR, AND, NOR 게이트를 구성하면 다음 그림과 같다.

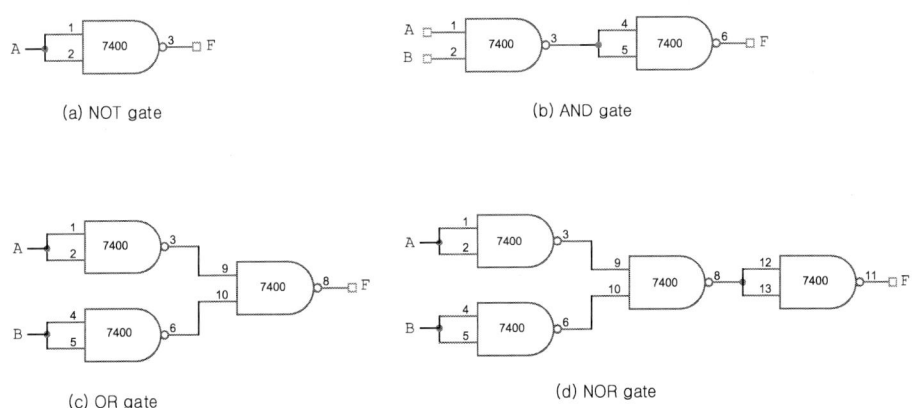

(a) NOT gate

(b) AND gate

(c) OR gate

(d) NOR gate

그림 2-4 NAND 게이트만을 이용한 기본 논리회로

(1) 그림 2-5와 같이 회로를 구성하고 각 지점(U, V, W, F1, F2)의 전압을 측정
하여 표 2-1에 기록하고 논리식을 작성하여 출력 F1과 F2를 비교하시오.

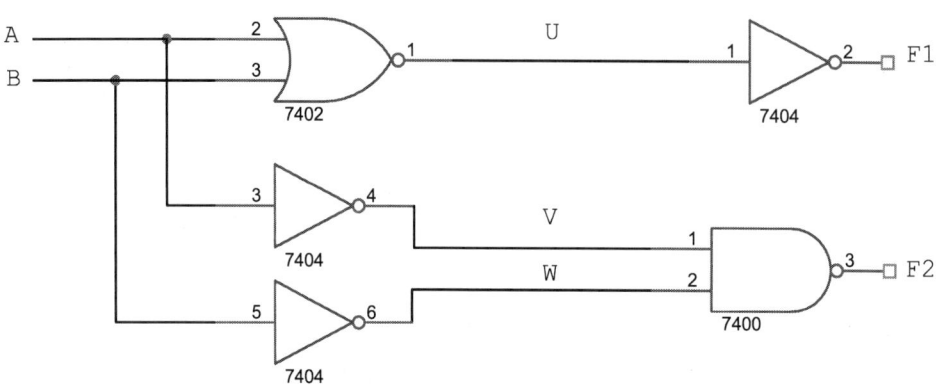

그림 2-5

표 2-1

A	B	U	V	W	F1	F2
0	0					
0	1					
1	0					
1	1					

■ 논리식

F1 =

F2 =

(2) 그림 2-6의 회로를 구성하고 각 지점(U, V, W, F1, F2)의 전압을 측정하여
　　표 2-2에 기록하고 논리식을 작성하여 출력 F1과 F2를 비교하시오.

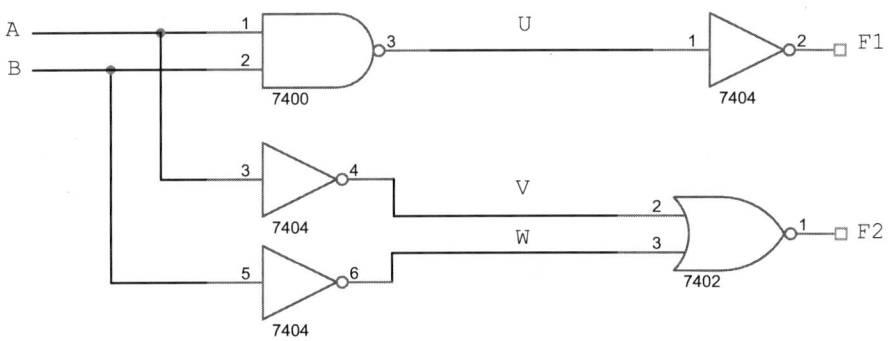

그림 2-6

표 2-2

A	B	U	V	W	F1	F2
0	0					
0	1					
1	0					
1	1					

■ 논리식

F1 =

F2 =

(3) 그림 2－7의 회로를 구성하고 출력전압을 측정하여 표 2－3에 기록하고 논리
식을 작성하여 출력 F1과 F2를 비교하시오.

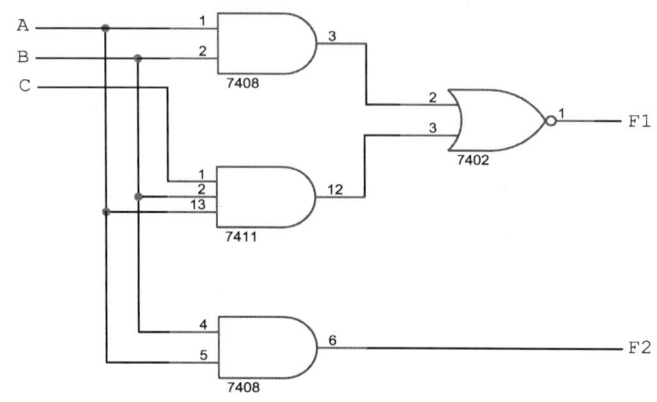

그림 2－7

표 2－3

A	B	C	F1	F2
0	0	0		
0	0	1		
0	1	0		
0	1	1		
1	0	0		
1	0	1		
1	1	0		
1	1	1		

■ 논리식

F1 =

F2 =

(4) 그림 2−8의 회로를 구성하고 출력전압을 측정하여 표 2−4에 기록하고 논리식을 작성하여 출력 F1과 F2를 비교하시오.

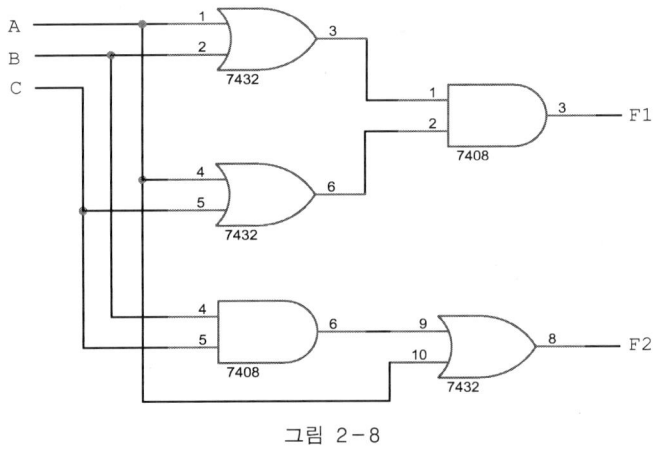

그림 2−8

표 2−4

A	B	C	F1	F2
0	0	0		
0	0	1		
0	1	0		
0	1	1		
1	0	0		
1	0	1		
1	1	0		
1	1	1		

■ 논리식

F1 =

F2 =

(5) 그림 2-9의 회로를 구성하고 출력전압을 측정하여 표 2-5에 기록하고 논리식을 작성하여 출력 F1, $\overline{F1}$ 과 F2를 비교하시오.

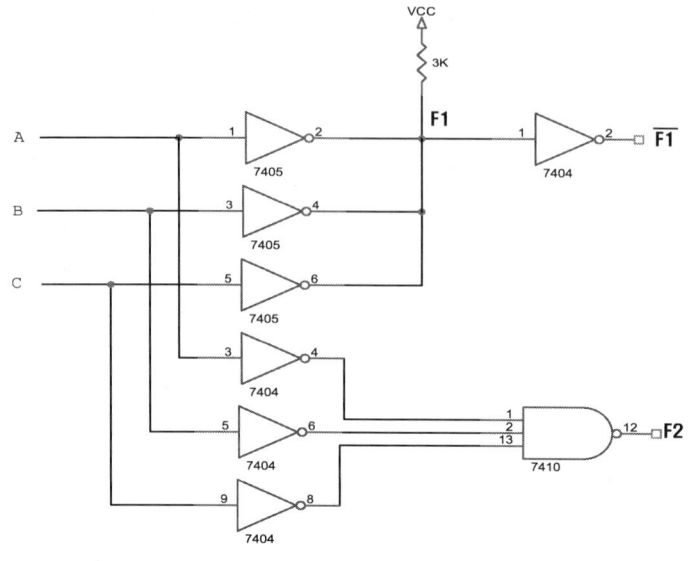

그림 2-9

표 2-5

A	B	C	F1	$\overline{F1}$	F2
0	0	0			
0	0	1			
0	1	0			
0	1	1			
1	0	0			
1	0	1			
1	1	0			
1	1	1			

■ 논리식

F1 =

$\overline{F1}$ =

F2 =

(6) 그림 2 − 10의 회로를 구성하고 각 지점(U, V, W, F1, F2)의 전압을 측정하여
표 2 − 6에 기록하고 논리식을 작성하여 상호 비교하시오.

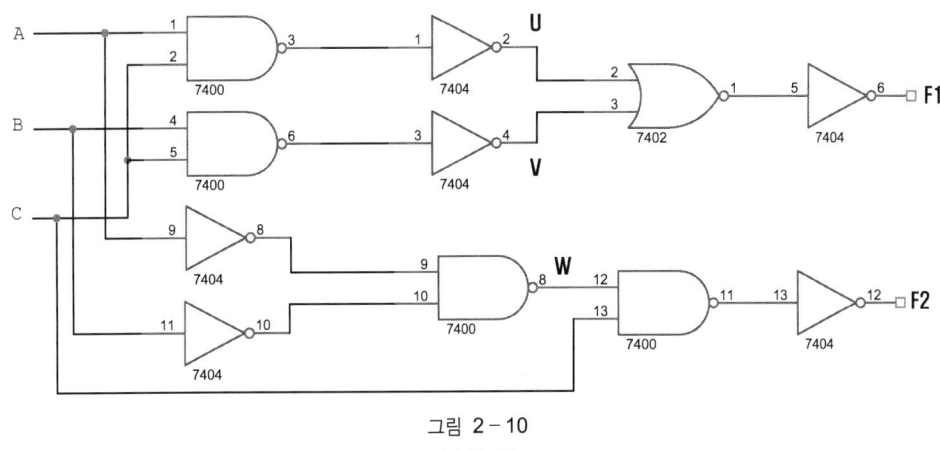

그림 2 − 10

표 2 − 6

A	B	C	U	V	W	F1	F2
0	0	0					
0	0	1					
0	1	0					
0	1	1					
1	0	0					
1	0	1					
1	1	0					
1	1	1					

■ 논리식

F1 =

F2 =

표 2-1

A	B	U	V	W	F1	F2
0	0					
0	1					
1	0					
1	1					

표 2-2

A	B	U	V	W	F1	F2
0	0					
0	1					
1	0					
1	1					

표 2-3

A	B	C	F1	F2
0	0	0		
0	0	1		
0	1	0		
0	1	1		
1	0	0		
1	0	1		
1	1	0		
1	1	1		

표 2-4

A	B	C	F1	F2
0	0	0		
0	0	1		
0	1	0		
0	1	1		
1	0	0		
1	0	1		
1	1	0		
1	1	1		

표 2-5

A	B	C	F1	$\overline{F1}$	F2
0	0	0			
0	0	1			
0	1	0			
0	1	1			
1	0	0			
1	0	1			
1	1	0			
1	1	1			

표 2-6

A	B	C	U	V	W	F1	F2
0	0	0					
0	0	1					
0	1	0					
0	1	1					
1	0	0					
1	0	1					
1	1	0					
1	1	1					

드리블이 잘 나갈 때 **어디선가** 들어오는 **태클**을 **대비**하라!

가산기와 감산기

실험목적

- 반가산기와 전가산기의 기본 원리를 이해하도록 한다.
- 반감산기와 전감산기의 기본 원리를 이해하도록 한다.
- 가산기 및 감산기를 이용하여 논리회로를 구성하는 능력을 키운다.
- 가산기를 이용한 가감산기를 이해하도록 한다.

실험기기 및 재료

구분	품명	규격	수량	비고
기기	논리회로 실험장치		1	
	회로시험기		1	
재료	NOT 게이트	IC 74LS04	1	
	NAND 게이트	IC 74LS00	1	
	NOR 게이트	IC 74LS02	1	
	EX — OR 게이트	IC 74LS86	1	
	LED		2	
	저항	330 Ω	2	1/4W
	점퍼선		약간	

실험3 가산기와 감산기 | 53

1. 반가산기(Half Adder)

두 개의 비트를 더하는 경우는 $0+0=0$, $0+1=1$, $1+0=1$, $1+1=10$ 등 4가지가 있다. 이 중 $1+1=10$의 경우는 합(sum: 0)과 캐리(carry: 1)가 발생됨을 알 수 있다. 이와 같은 과정을 수행하는 장치를 반가산기(Half Adder)라고 하며 진리표는 표 3－1과 같다.

표 3－1 반가산기의 진리표

입력		출력	
A	B	C	S
0	0	0	0
0	1	0	1
1	0	0	1
1	1	1	0

그림 3－1 반가산기의 기호

위의 표에서 합(S)과 캐리(C)에 대한 논리식을 구하면 다음과 같다.

$$S = \overline{A}B + A\overline{B} = A \oplus B$$

$$C = A \cdot B$$

위의 논리식을 이용하여 반가산기를 구성하면 그림 3－2와 같이 된다.

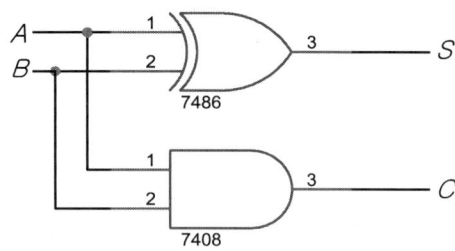

그림 3－2 반가산기 회로

2. 전가산기(Full Adder)

	$C_{n+1}=1$	$C_n=1$	$C_{n-1}=1$	
	1	1	1	.... (A)
+	1	0	1	.... (B)
1	1	0	0	
C_{n+1}	S_{n+1}	S_n	S_{n-1}	

3개의 비트를 가산하는 경우를 생각해 보자. A＝111과 B＝101을 더하는 경우 n번째 자리에서 더해지는 과정을 보면 n－1번째 자리에서 발생된 캐리($C_{n}-1$)과 A(1), B(0)의 세 수가 더해져서 합(Sn)은 0이 되고, 다시 이 자리에서 캐리(Cn)가 발생하여 다음 자리에 더해지게 된다. 이러한 과정을 수행하는 장치를 전가산기(Full Adder)라고 하며 이의 진리표와 기호는 아래와 같다.

표 3 - 2 반가산기의 진리표

입력			출력	
A_n	B_n	C_{n-1}	C_n	S_n
0	0	0	0	0
0	0	1	0	1
0	1	0	0	1
0	1	1	1	0
1	0	0	0	1
1	0	1	1	0
1	1	0	1	0
1	1	1	1	1

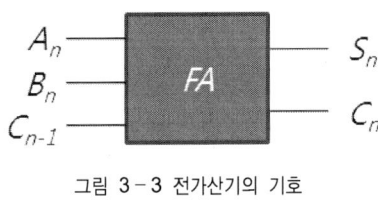

그림 3 - 3 전가산기의 기호

위의 진리표에 의해 합(S_n)과 캐리(C_n)를 구하면 다음과 같다.

$$S_n = \overline{A_n}\,\overline{B_n}\,C_{n-1} + \overline{A_n}\,B_n\,\overline{C_{n-1}} + A_n\,\overline{B_n}\,\overline{C_{n-1}} + A_n B_n C_{n-1}$$
$$= A_n \oplus B_n \oplus C_{n-1}$$

$$C_n = \overline{A_n}\,B_n\,C_{n-1} + A_n\,\overline{B_n}\,C_{n-1} + A_n\,B_n\,\overline{C_{n-1}} + A_n B_n C_{n-1}$$
$$= C_{n-1}(A_n \oplus B_n) + A_n B_n$$

위의 논리식을 이용하여 전가산기를 구성하면 그림 3 - 4와 같이 된다.

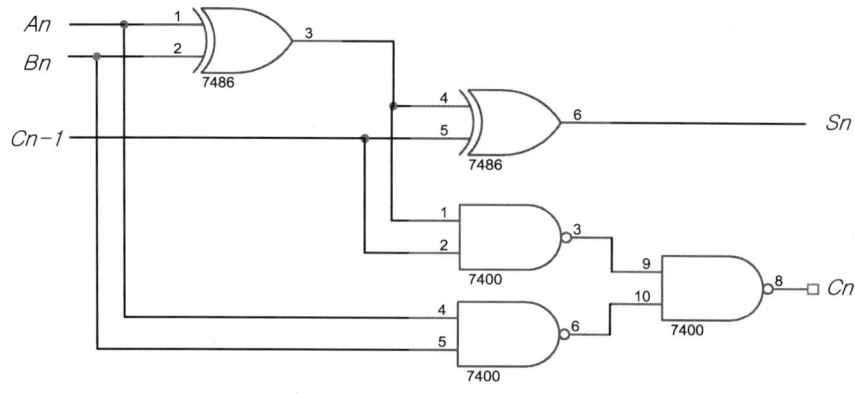

그림 3-4 전가산기 회로

그림 3-5와 같이 하면 반가산기 두 개를 이용하여 전가산기를 구성할 수도 있다.

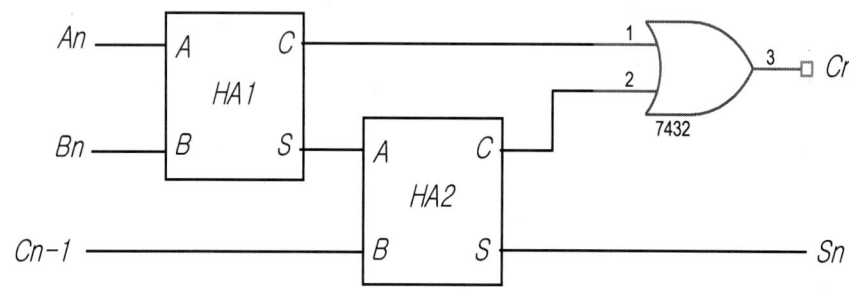

그림 3-5 반가산기를 이용한 전가산기의 구성

3. 반감산기(Half Subtracter)

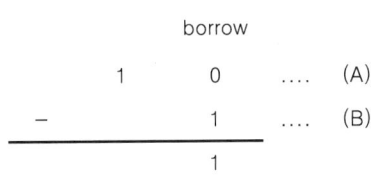

두 수를 빼기(A – B)를 하는 경우를 생각해 보자. 옆의 경우처럼 빼기를 하기 위해서 앞자리에서 1을 빌려온(borrow) 다음 B를 빼면 차(difference)는 1이 됨을 알 수 있다. 이와 같은 과정을 수행하는 회로를 반감산기라 하고 반감산기 회로의 입력을 A와 B, 출력을 d, 자리 빌림을 b라 하면 표 3 – 3과 같은 진리표를 얻을 수 있다.

표 3 – 3 반감산기의 진리표

입력		출력	
A	B	b	d
0	0	0	0
0	1	1	1
1	0	0	1
1	1	0	0

이 진리표에서 다음과 같이 출력이 구해진다.

$$d = \overline{A}B + A\overline{B} = A \oplus B$$

$$b = \overline{A}B$$

위의 논리식을 이용하여 반감산기를 구성하면 그림 3 – 6과 같다.

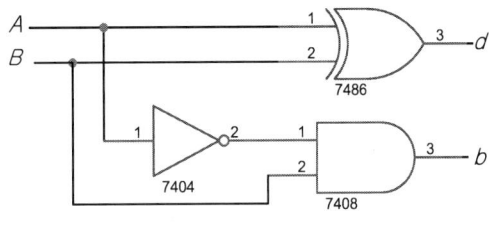

그림 3 – 6 반감산기 회로

4. 전감산기(Full Subtracter)

$b_{n+1} = 1$ $b_n = 1$

1	1	0	.... (A)
− 0	1	1	.... (B)
0	1	1	

3개의 비트를 빼기(A − B)하는 경우를 생각해 보자. 옆의 경우에서 n번째 자리에서 수행되는 과정을 살펴보면 n번째 자리의 감산을 수행하기 전에 이미 n −

1번째 자리에 1을 빌려준 상태가 되어 있고 현재의 An = 0이므로 다시 n + 1번째 자리에서 1을 빌려야 된다. 이처럼 An − Bn − bn에 대해 출력이 bn + 1과 d로 표시되는 회로를 전감산기라고 한다. 이의 진리표는 아래와 같다.

표 3−4 반감산기의 진리표

입력			출력	
A_n	B_n	b_n	b_{n+1}	d
0	0	0	0	0
0	0	1	1	1
0	1	0	1	1
0	1	1	1	0
1	0	0	0	1
1	0	1	0	0
1	1	0	0	0
1	1	1	1	1

이 진리표에서 다음과 같은 출력을 구할 수 있다.

$$d = \overline{A_n}\,\overline{B_n}b_n + \overline{A_n}B_n\overline{b_n} + A_n\overline{B_n}\,\overline{b_n} + A_nB_nb_n$$

$$= A_n \oplus B_n \oplus b_n$$

$$b_{n+1} = \overline{A_n}\,\overline{B_n}b_n + \overline{A_n}B_n\overline{b_n} + \overline{A_n}B_nb_n + A_nB_nb_n$$

$$= \overline{A_n}(B_n \oplus b_n) + B_nb_n$$

위의 논리식을 이용하여 전감산기를 구성하면 그림 3-7과 같다.

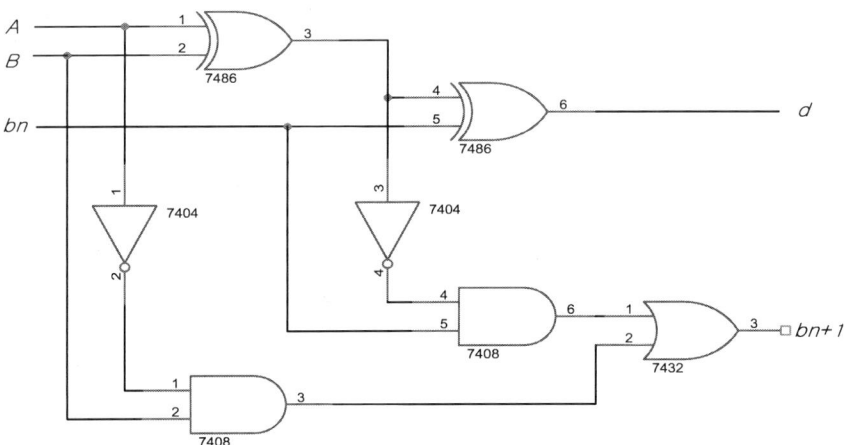

그림 3-7 전감산기 회로

$$b_{n+1} = A_n{'}B_n + A_n{'}B_n{'}b_n + A_nB_nb_n$$

$$= A_n{'}B_n + b_n(A_n{'}B_n{'} + A_nB_n)$$

$$= A_n{'}B_n + b_n(A_n \oplus B_n){'}$$

전감산기는 반감산기 두 개를 이용하여 아래 그림과 같이 구성할 수도 있다.

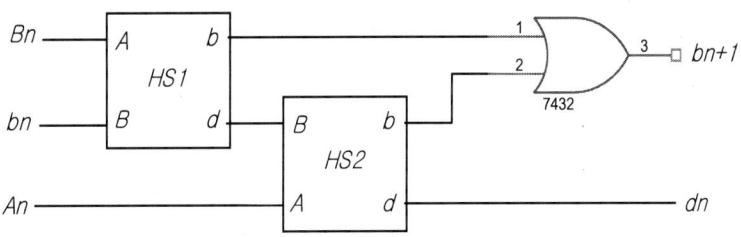

그림 3-8 반감산기를 이용한 전감산기의 구성

(1) 그림 3-9와 같이 회로를 구성하고 출력 상태를 측정하여 표 3-5에 기록하고 논리식을 작성하시오.

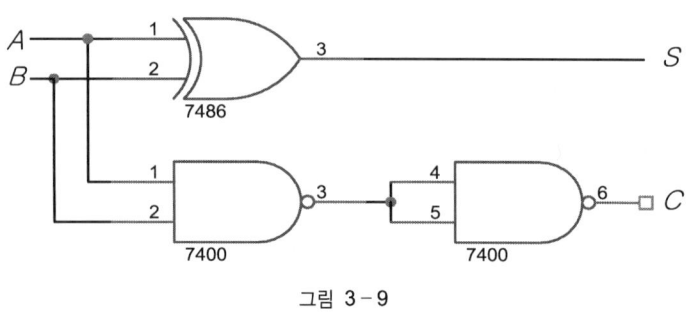

그림 3-9

표 3-5

입력		출력	
A	B	C	S
0	0		
0	1		
1	0		
1	1		

■ 논리식

(1) C =

(2) S =

(3) 어떤 회로인가?

(2) 그림 3 – 10과 같이 회로를 구성하고 출력 상태를 측정하여 표 3 – 6에 기록하고 논리식을 작성하시오.

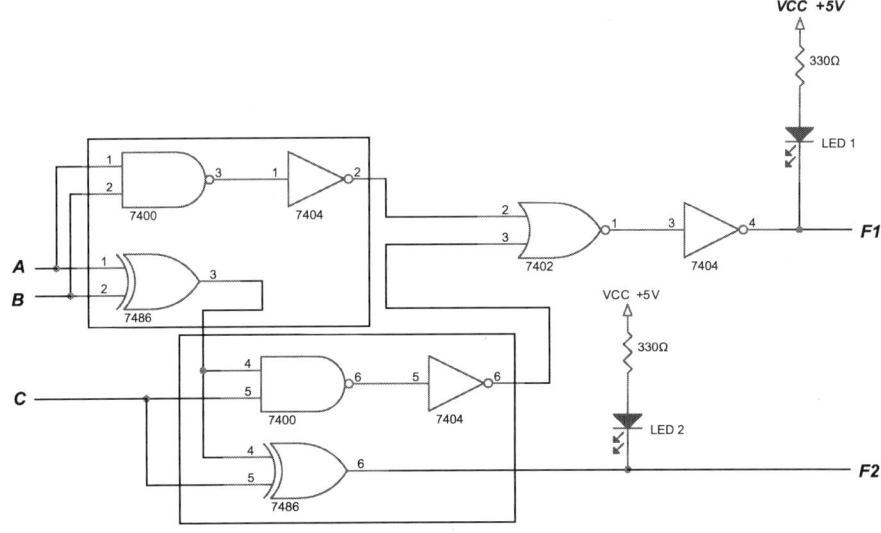

그림 3 – 10

표 3 – 6

입력			출력	
A	B	C	F1	F2
0	0	0		
0	0	1		
0	1	0		
0	1	1		
1	0	0		
1	0	1		
1	1	0		
1	1	1		

■ 논리식

(1) F1 =

(2) F2 =

(3) 어떤 회로인가?

(3) 그림 3-11과 같이 회로를 구성하고 출력 상태를 측정하여 표 3-7에 기록하고 논리식을 작성하시오.

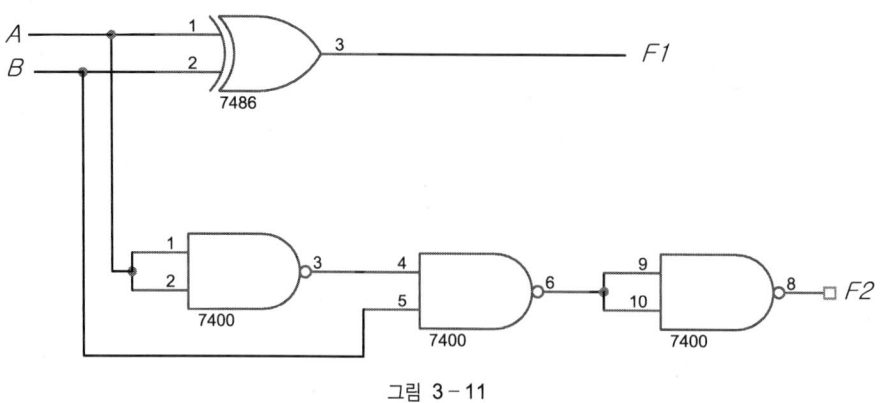

그림 3-11

표 3-7

입력		출력	
A	B	F1	F2
0	0		
0	1		
1	0		
1	1		

■ 논리식

(1) F1 =

(2) F2 =

(3) 어떤 회로인가?

(3) 그림 3 – 12와 같이 회로를 구성하고 출력 상태를 측정하여 표 3 – 8에 기록하고 논리식을 작성하시오.

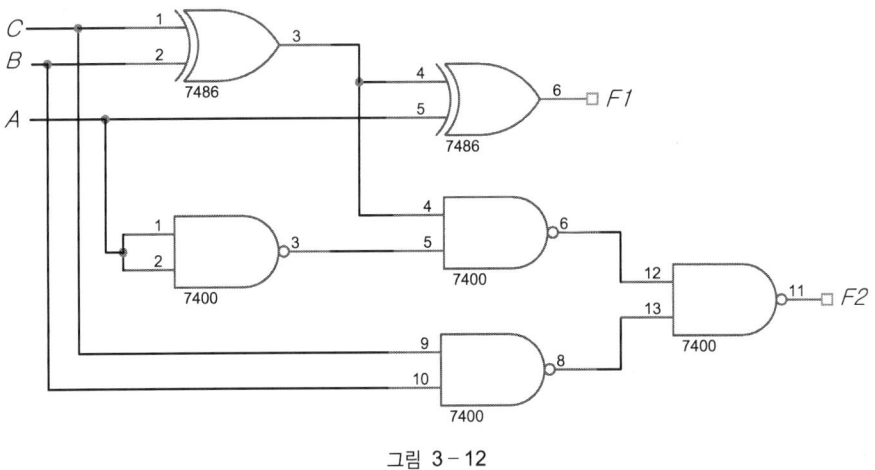

그림 3 – 12

표 3 – 8

입력			출력	
A	B	C	F1	F2
0	0	0		
0	0	1		
0	1	0		
0	1	1		
1	0	0		
1	0	1		
1	1	0		
1	1	1		

■ 논리식

(1) F1 =

(2) F2 =

(3) 어떤 회로인가?

표 3-5

입력		출력	
A	B	C	S
0	0		
0	1		
1	0		
1	1		

■ 논리식

(1) C =

(2) S =

(3) 어떤 회로인가?

표 3-6

입력			출력	
A	B	C	F1	F2
0	0	0		
0	0	1		
0	1	0		
0	1	1		
1	0	0		
1	0	1		
1	1	0		
1	1	1		

■ 논리식

(1) F1 =

(2) F2 =

(3) 어떤 회로인가?

표 3-7

입력		출력	
A	B	F1	F2
0	0		
0	1		
1	0		
1	1		

■ 논리식

(1) F1 =

(2) F2 =

(3) 어떤 회로인가?

표 3-8

입력			출력	
A	B	C	F1	F2
0	0	0		
0	0	1		
0	1	0		
0	1	1		
1	0	0		
1	0	1		
1	1	0		
1	1	1		

■ 논리식

(1) F1 =

(2) F2 =

(3) 어떤 회로인가?

실험 4   디코더와 엔코더

실험목적

● 디코더의 동작원리와 회로 구성방법을 이해하도록 한다.
● 엔코더의 원리를 이해하고 이를 이용, 각종 코드를 만들 수 있도록 한다.
● 엔코더로 코드화된 신호를 해석할 수 있도록 한다.

실험기기 및 재료

구분	품명	규격	수량	비고
기기	논리회로 실험장치		1	
	회로시험기		1	
재료	AND 게이트	IC 74LS08	1	
	OR 게이트	IC 74LS32	1	
	Decoder(BCD to 7 ─ Seg)	IC 74LS47	1	
	저항	330 Ω	7	1/4W
	FND 500		1	common anode
	점퍼선		약간	

1. 디코더(Decoder)

디코더란 입력선에 나타나는 n비트 2진 코드를 최대 2^n개의 서로 다른 정보로 해독(decoding)해 주는 회로를 말한다. 따라서 디코더는 해독기(解讀器)라고도 한다. 만일 n비트의 디코더 정보가 모두 사용되지 않거나 리던던시(Redundancy) 조합을 가진다면 디코더의 출력 수는 2^n개보다 적게 된다. 표 4-1은 BCD 코드를 10진으로 해독해 주는 BCD to 10진 디코더의 진리표를 나타낸다. 여기서 BCD 코드를 한 조합에 대해서 D0, D1, ……, D9 중 하나만 동작할 수 있도록 디코딩하면

$$D_0 = \overline{D}\,\overline{C}\overline{B}\overline{A}$$

$$D_1 = \overline{D}\,\overline{C}\overline{B}A$$

$$\cdots\cdots\cdots\cdots\cdots$$
$$D_9 = D\,\overline{C}\overline{B}A$$

가 되므로 그림 4-1과 같이 BCD to 10진 디코더 회로를 구성할 수 있다.

표 4-1 BCD to 10진 디코더 진리표

입력(BCD 코드)				출력(10진수)									
D	C	B	A	D_0	D_1	D_2	D_3	D_4	D_5	D_6	D_7	D_8	D_9
0	0	0	0	1	0	0	0	0	0	0	0	0	0
0	0	0	1	0	1	0	0	0	0	0	0	0	0
0	0	1	0	0	0	1	0	0	0	0	0	0	0
0	0	1	1	0	0	0	1	0	0	0	0	0	0
0	1	0	0	0	0	0	0	1	0	0	0	0	0
0	1	0	1	0	0	0	0	0	1	0	0	0	0
0	1	1	0	0	0	0	0	0	0	1	0	0	0
0	1	1	1	0	0	0	0	0	0	0	1	0	0
1	0	0	0	0	0	0	0	0	0	0	0	1	0
1	0	0	1	0	0	0	0	0	0	0	0	0	1

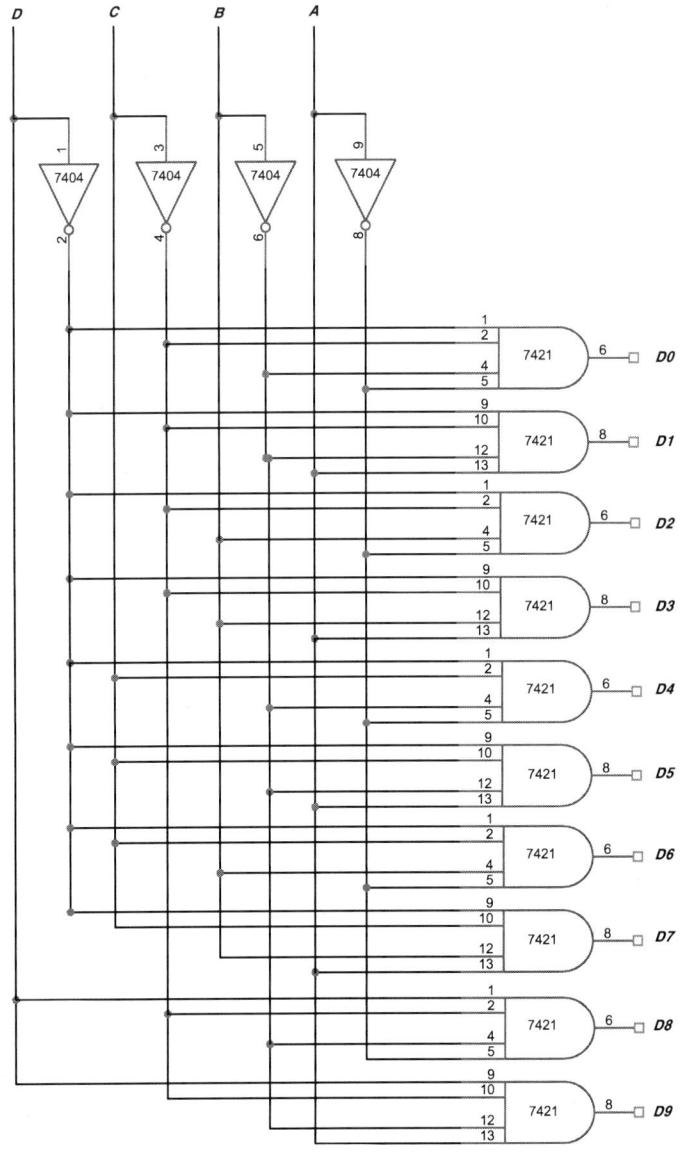

그림 4-1 BCD to 10진 디코더 회로

위와 같은 회로를 하나의 IC로 실현한 것이 있는데 TTL IC로서 대표적인 BCD to 10진 디코더(BCD to Decoder)는 74LS42가 있다.

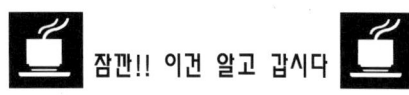 잠깐!! 이건 알고 갑시다

BCD to 7 Segment Decoder

디코더의 또 다른 응용 형태로 BCD 코드를 입력받아 대응된 10진수의 값을 7 - 세그먼트에 디스플레이해 주는 디코더가 있는데 이것이 바로 BCD to 7 Segment Decoder이다.

(1) 7 - Segment

LED를 숫자의 형태로 배열하여 입력이 있을 때 그에 대응한 숫자를 표시하도록 점등되는 것을 7 - Segment라고 한다. 모양은 아래 그림과 같다.

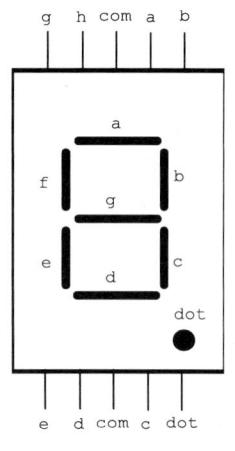

그림 4 - 2 7 - Segment

7 - Segment는 애노드(anode) 공통형과 캐소드(Cathode) 공통형이 있는데 전자는 입력으로 Low신호를, 후자는 High신호를 사용한다. 또한 LED는 정격전압이 2V, 전류는 약 9mA가 흐르므로 5V 전압을 사용하는 경우 전류 제한용 저항 약 330Ω을, 9V를 사용하는 경우 약 1㏀의 저항을 외부에 연결하여 사용해야 한다. 자세한 사항은 '실험 14. 카운터의 응용'에서 다루기로 하겠다.

(2) BCD to 7 - Segment Decoder

대표적인 BCD to 7 - Segment Decoder로는 74LS47이 있으며 이 디코더를 이용하여 LED와 접속할 때는 그림 4 - 3과 같이 하면 된다.

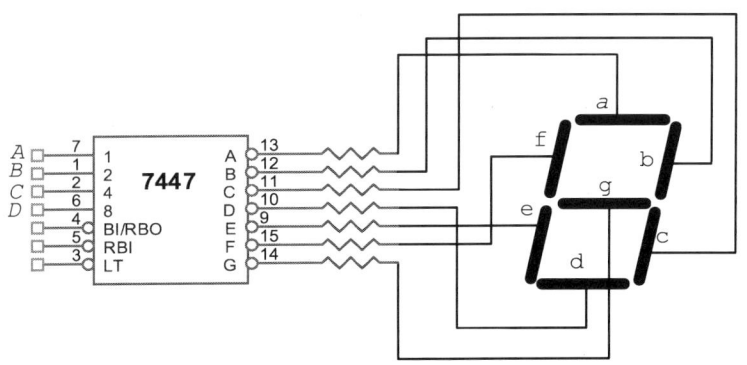

그림 4 - 3 디코더와 7 - Segment의 접속회로

2. 엔코더(Encoder)

엔코더는 디코더의 역연산을 하는 조합 회로로 2n개의 입력과 n개의 출력정보로 내보내는 장치이다. 따라서 엔코더를 이용하면 우리가 요구하는 어떠한 종류의 코드도 만들 수 있다. 10진수를 BCD 코드로 변환하는 엔코더(Decimal to BCD encoder)를 예를 들어 설명하자. 입력은 십진수 0～9까지이며 출력은 BCD 코드에 대응한 4개가 된다. 이의 진리표는 표 4 - 2와 같다.

표 4 - 2 10진 to BCD 엔코더의 진리표

입력	출력			
십진수	D	C	B	A
0	0	0	0	0
1	0	0	0	1
2	0	0	1	0
3	0	0	1	1
4	0	1	0	0
5	0	1	0	1
6	0	1	1	0
7	0	1	1	1
8	1	0	0	0
9	1	0	0	1

이 진리표에서 출력 D, C, B, A를 십진수를 입력으로 하여 논리식을 구하면 다음과 같다.

$D = 8 + 9$

$C = 4 + 5 + 6 + 7$

$B = 2 + 3 + 6 + 7$

$A = 1 + 3 + 5 + 7 + 9$

위의 논리식을 이용하여 논리회로를 만들면 그림 4-4와 같다.

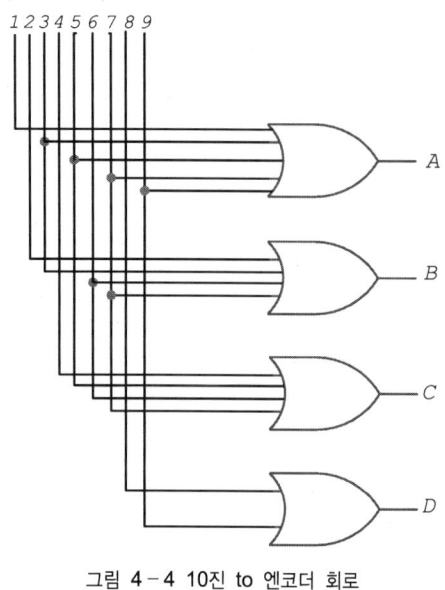

그림 4-4 10진 to 엔코더 회로

 잠깐!! 이건 알고 갑시다

10진-BCD 우선순위 엔코더

엔코더는 주어진 시간에 하나의 입력만을 받아들이는데 동시에 2개 이상의 입력
이 들어오면 에러가 발생하게 된다. 따라서 엔코더에 변형을 가하여 2개 이상의 입
력이 들어오더라도 입력의 10진수 중 가장 높은 수에 대응하는 BCD 코드가 출력
되도록 하는 우선순위(Priority)를 부여하여 에러를 방지하게 된다. 이런 엔코더를
우선순위 엔코더(Priority encoder)라 한다. 예를 들어 3과 6이 동시에 입력되면 6이
우선하여 BCD 출력은 110이 된다. 이런 회로를 실현하는 대표적인 IC로는
74LS147이 있다.

(1) 그림 4 − 5와 같이 회로를 구성하고 출력 상태를 측정하여 표 4 − 3에 기록하고 논리식을 작성하시오.

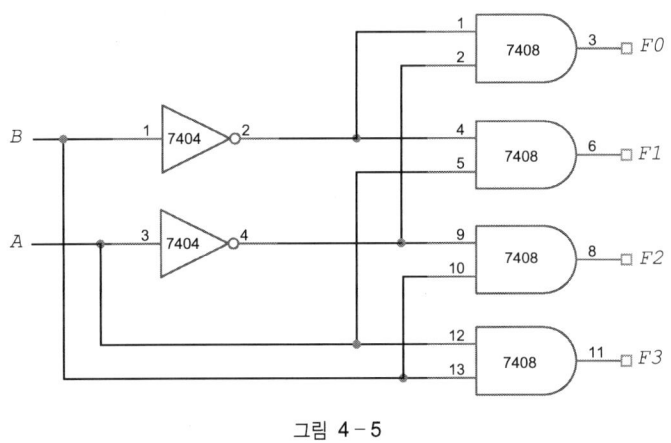

그림 4 − 5

표 4 − 3

입력		출력			
B	A	F0	F1	F2	F3

■ 논리식

(1) F0 =

(2) F1 =

(3) F2 =

(4) F3 =

(5) 어떤 회로인가?

(2) 그림 4 - 6과 같이 회로를 구성하고 출력 상태를 측정하여 표 4 - 4에 기록하고
   논리식을 작성하시오.

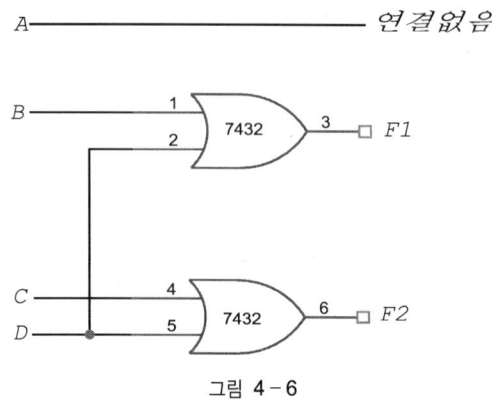

그림 4 - 6

표 4 - 4

입력				출력	
A	B	C	D	F1	F2

■ 논리식

(1) F1 =

(2) F2 =

(3) 어떤 회로인가?

(3) 그림 4 − 7과 같이 회로를 구성하고 출력 상태를 측정하여 표 4 − 5에 기록하고
논리식을 작성하시오.

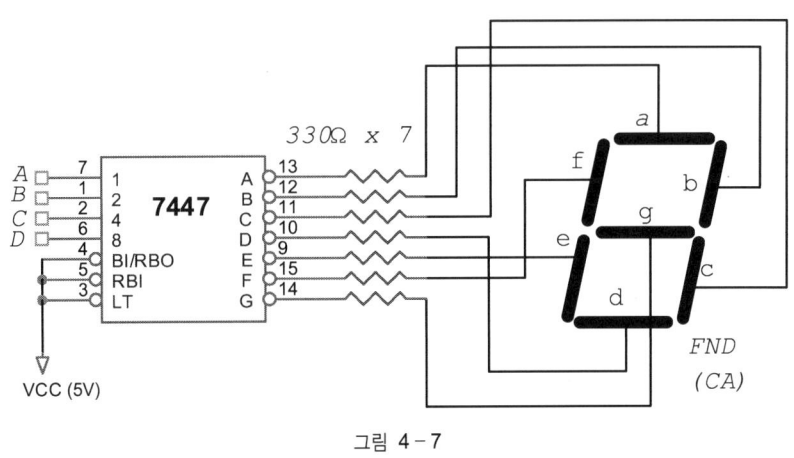

그림 4 − 7

표 4 − 5

입력				출력(FND)							
D	C	B	A	a	b	c	d	e	f	g	표시된 형태
0	0	0	0								
0	0	0	1								
0	0	1	0								
0	0	1	1								
0	1	0	0								
0	1	0	1								
0	1	1	0								
0	1	1	1								
1	0	0	0								
1	0	0	1								

실험결과 Report		학과명	학번	성명
실험 4	디코더와 엔코더			

표 4 - 3

입력		출력			
B	A	F0	F1	F2	F3

■ 논리식

(1) F0 =

(2) F1 =

(3) F2 =

(4) F3 =

(5) 어떤 회로인가?

표 4-4

입력				출력	
A	B	C	D	F1	F2

■ 논리식

(1) F1 =

(2) F2 =

(3) 어떤 회로인가?

표 4-5

입력				출력(FND)							
D	C	B	A	a	b	c	d	e	f	g	표시된 형태
0	0	0	0								
0	0	0	1								
0	0	1	0								
0	0	1	1								
0	1	0	0								
0	1	0	1								
0	1	1	0								
0	1	1	1								
1	0	0	0								
1	0	0	1								

실험 5 | 멀티플렉서와 디멀티플렉서

실험목적

- 멀티플렉서의 구성방법 및 동작원리를 이해하도록 한다.
- 멀티플렉서를 이용하여 논리회로의 간소화를 실현할 수 있도록 한다.
- 디멀티플렉서의 등가회로를 그리고 동작원리를 이해하도록 한다.
- 멀티플렉서와 디멀티플렉서를 이용한 응용회로를 설계할 수 있도록 한다.

실험기기 및 재료

구분	품명	규격	수량	비고
기기	논리회로 실험장치		1	
	회로시험기		1	
재료	NAND 게이트	IC 74LS20	3	4입력 NAND
	NAND 게이트	IC 74LS10	2	3입력 NAND
	NOT 게이트	IC 74LS04	1	
	점퍼선		약간	

1. 멀티플렉서(Multiplexer)

  멀티플렉싱(Multiplexing)이란 '**많은 수의 정보를 적은 수의 채널이나 선들을 통하여 전송하는 것**'으로 데이터 선택기(Data Selector)라고도 한다. 현재 사용되고 있는 많은 수의 통신장비들이 정보전송을 보다 효율적으로 수행하기 위해 멀티플렉싱 개념을 사용하고 있다. 멀티플렉싱을 해 주는 멀티플렉서는 여러 개의 입력신호 중 하나의 입력만을 선택하여 하나의 출력선을 통해 내보내는 회로를 말한다. 그림 5-1은 멀티플렉서에 대한 기본적인 원리를 설명해 준다.

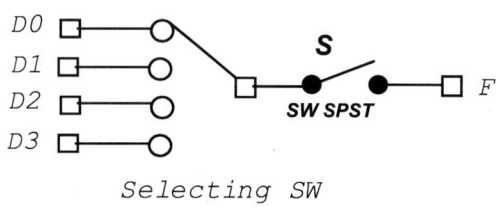

Selecting SW

그림 5-1 멀티플렉서 개념 스위치

  그림 5-2와 같은 회로에서 D_0, D_1, D_2, D_3의 입력신호를 선택선 A, B에 의해 1개만 선택하여 전송하는 경우를 보면 인에이블(Enable)선 S=1인 경우는 모든 AND 게이트의 출력이 '0'이 되므로 어떤 입력신호도 선택할 수 없게 되고 S=0이고 A=0, B=0이면 D_0를, A=0, B=1이면 D1을, A=1, B=0이면 D_2를, A=1, B=1이면 D_3를 선택하여 출력하게 된다.

  따라서 멀티플렉서의 출력은

$$F = \overline{S}(D_0\overline{A}\,\overline{B} + D_1\overline{A}B + D_2A\overline{B} + D_3AB)$$

로 표시할 수 있다.

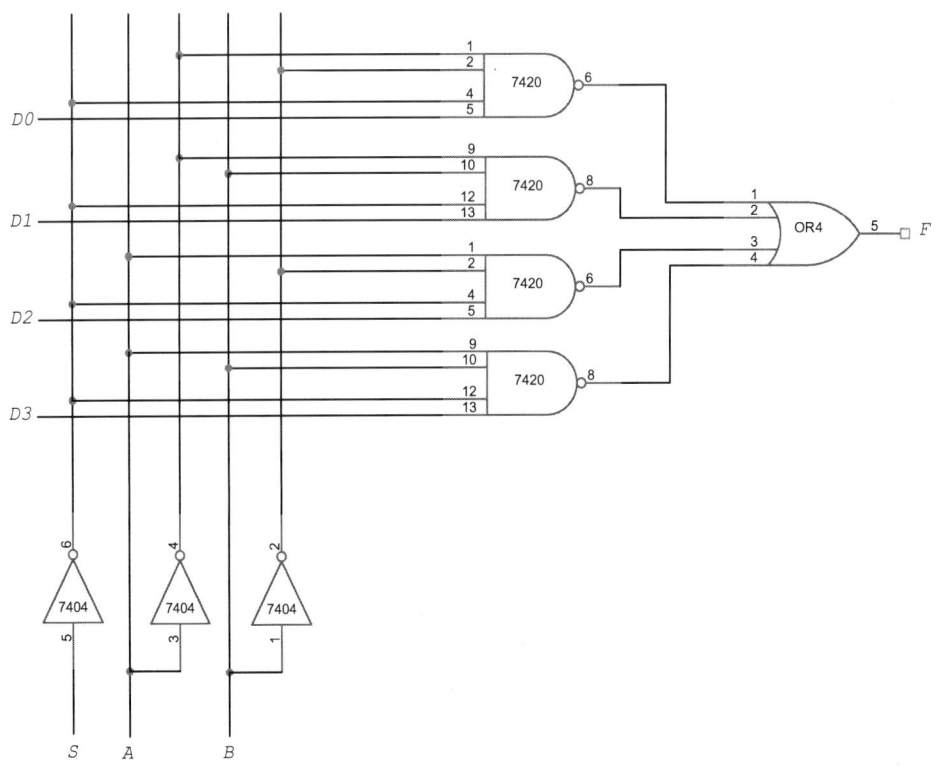

그림 5 - 2 멀티플렉서 논리회로

위의 멀티플렉서의 논리회로는 그림 5 - 3과 같이 블록도로 나타낼 수 있으며 함수표는 표 5 - 1과 같다.

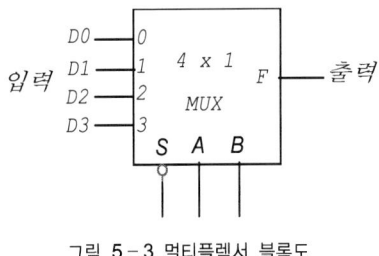

그림 5 - 3 멀티플렉서 블록도

표 5 - 1 멀티플렉서 함수표

입력			출력
S	A	B	F
0	0	0	D0
0	0	1	D1
0	1	0	D2
0	1	1	D3

2. 멀티플렉서를 이용한 논리회로의 구현

멀티플렉서를 이용하면 논리회로를 구현하기가 매우 쉽다. 다음의 예를 살펴보도록 하자.

$F = A \oplus B = \overline{A}B + A\overline{B}$의 논리함수의 함수표는 표 5-2와 같고 이것은 그림 5-4와 같이 간단히 실현할 수 있다.

표 5-2 함수표

입력			출력
S	A	B	F
0	0	0	0
0	0	1	1
0	1	0	1
0	1	1	0

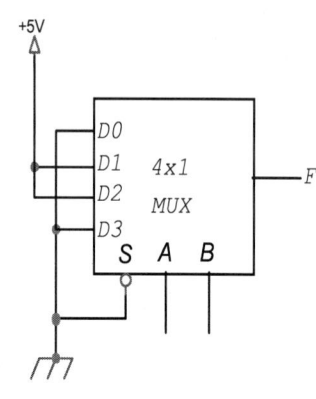

그림 5-4 멀티플렉서에 의한
논리회로

위의 예에서도 알 수 있듯이 멀티플렉서는 OR 게이트와 디코더의 결합형태와 같다고 생각할 수 있다. MUX에서 선택선(A, B)을 제어함으로써 입력변수들을 제어할 수 있으며 이때 함수의 변수들은 선택선에 연결된다. 수행될 함수에 포함되어 있는 입력변수들은 대응되는 입력들을 '1'로 만듦으로써 선택되고 함수에 포함되어 있지 않은 변수들은 대응되는 입력선을 '0'으로 만들면 된다. 다음에 일반적인 예를 들어 설명해 보자.

만일 n+1개의 변수들로 이루어진 함수가 있다면 이 변수들 중 n개를 선택해서 MUX의 선택선으로 사용하고, 나머지 1개는 MUX의 입력변수로 사용한다. 변수 A, B, C로 이루어진 다음의 함수가 있을 때 이를 MUX로 실현해 보자.

$$F(A, B, C) = \Sigma(1, 3, 5, 6) = \overline{A}\,\overline{B}C + \overline{A}BC + + A\overline{B}C + AB\overline{C}$$

위 함수의 진리표는 표 5-3과 같이 된다. 입력변수 A, B, C 중 A를 MUX의 입력으로 사용하고 나머지 B와 C는 선택선으로 사용한다고 하면 이 함수의 실현표는 아래 표 5-4와 같이 된다.

표 5-3 진리표

민텀	입력			출력
	A	B	C	F
m_0	0	0	0	0
m_1	0	0	1	1
m_2	0	1	0	0
m_3	0	1	1	1
m_4	1	0	0	0
m_5	1	0	1	1
m_6	1	1	0	1
m_7	1	1	1	0

표 5-4 실현표

	D_0	D_1	D_2	D_3
A'	0	①	2	③
A	4	⑤	⑥	7
	0	1	A	A'

위의 함수는 그림 5-5와 같이 4×1 MUX로 실현할 수 있다. 변수 A를 MUX의 입력으로 사용하고 B와 C는 MUX의 선택선으로 사용하여 각각 S_1과 S_0에 연결한다. MUX의 입력들을 0, 1, A, A'이라 하자. BC=00일 때 MUX의 D_0가 선택되고 D_0=0이기 때문에 F=0이 된다. BC=01일 때는 MUX의 D_1이 선택되고 D_1=1이므로 F=1이 되고, BC=10일 때는 MUX의 D_2가 선택되고 D_2=A이므로 F는 A값을 출력하고, BC=01일 때는 MUX의 D_3가 선택되고 D_3=A'여서 F=A'값을 출력하게 된다.

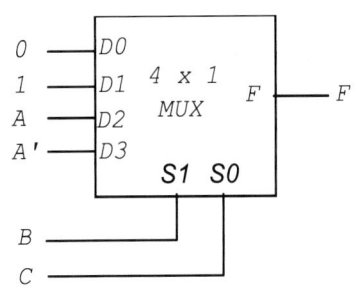

그림 5-5 멀티플렉서로 함수 실현

3. 디멀티플렉서(Demultiplexer)

디멀티플렉서는 멀티플렉서와 반대로 '**정보를 한선으로 받아서 2^n개의 출력선들 중 하나를 선택하여 받은 정보를 전송하는 회로**'이다. 디멀티플렉서는 n 개의 선택선에 의해 출력선이 선택된다. 디멀티플렉서는 인에이블 입력을 가지고 있는 디코더와 같다. 그림 5-6의 디코더에서 E가 입력선이고 A와 B가 선택선이라면 이것은 디멀티플렉서로 사용될 수 있다.

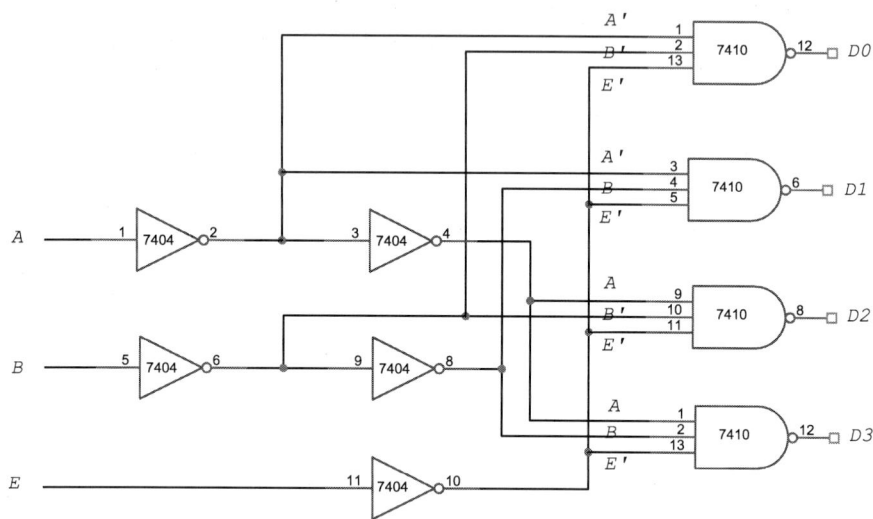

그림 5-6 인에이블 입력을 가진 **2x4** 디코더 회로

표 5-5 진리표

입력			출력			
E	A	B	D_0	D_1	D_2	D_3
1	x	x	1	1	1	1
0	0	0	0	1	1	1
0	0	1	1	0	1	1
0	1	0	1	1	0	1
0	1	1	1	1	1	0

인에이블 단자 E는 모든 출력 NAND 게이트와 연결되어 있기 때문에 E＝1일 경우에는 출력값이 항상 '1'이 된다. 위 회로에서 만일 인에이블 단자 E를 입력선으로, 변수 A, B를 선택선으로 사용한다면 입력 E의 값은 AB＝00일 때 D_0 출력에, AB＝01일 때는 D_1 출력에, AB＝10일 때는 D_2 출력에, AB＝11일 때는 D_3 출력에 나타나게 된다.

디코더의 입력선과 선택선을 서로 바꾸면 디멀티플렉서로 사용할 수가 있는데 디멀티플렉서는 다음 그림 5－7(b)와 같은 기호로 표시된다.

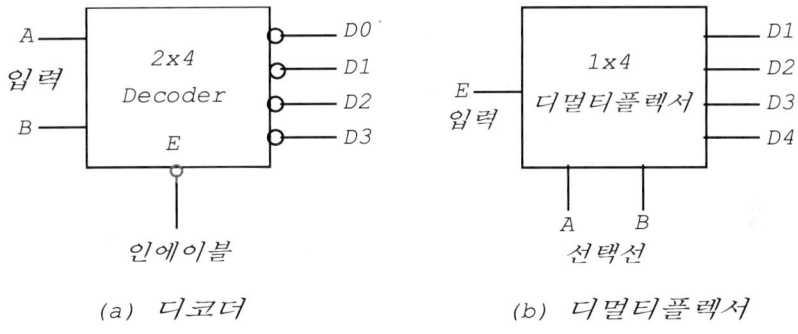

(a) 디코더   (b) 디멀티플렉서

그림 5－7 디코더와 디멀티플렉서 기호

(1) 그림 5-8과 같이 회로를 구성하고 출력 상태를 측정하여 표 5-6에 기록하고 논리식을 작성하시오(단 데이터의 전송은 D_0, D_1, D_2, D_3에 접속된 스위치를 각각 연속적으로 단속하여 시험하고 전송하지 않을 때는 스위치의 위치를 접지 측에 놓을 것).

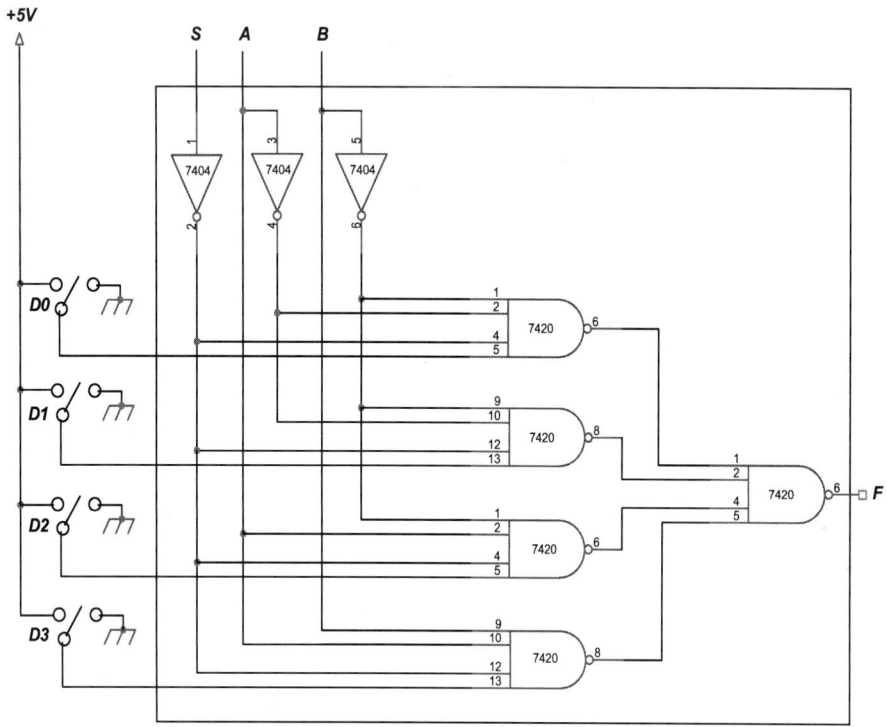

그림 5-8

표 5-6

입력			출력(F)			
S	A	B	D_0	D_1	D_2	D_3
0	0	0				
0	0	+5V				
0	+5V	0				
0	+5V	+5V				
+5V	0	0				
+5V	0	+5V				
+5V	+5V	0				
+5V	+5V	+5V				

※ 출력(F)의 표시는 데이터가 전송된 것에는 'O' 표를, 전송되지 않은 것에는 'X' 표를 하시오.

■ 논리식

(1) F =

(2) 어떤 회로인가?

(2) 실험(1)에서 구성한 그림 5-8의 회로를 그림 5-9와 같이 구성하여 출력 상
태를 측정하여 표 5-7에 기록하고 논리식을 작성하시오.

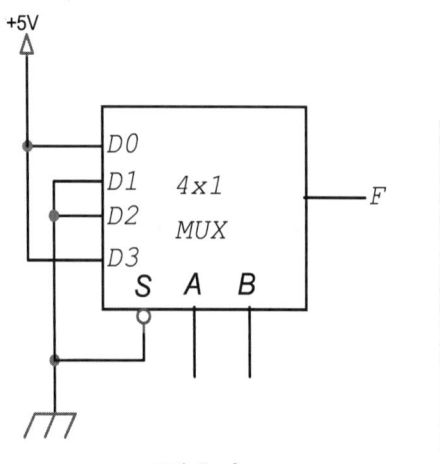

그림 5-9

표 5-7

입력		출력
A	B	F
0	0	
0	+5V	
+5V	0	
+5V	+5V	

■ 논리식

(1) F =

(2) 어떤 회로인가?

(3) 그림 5 - 10의 회로를 구성하고 출력 상태를 측정하여 표 5 - 8에 기록 후 논리
   식을 작성하시오.

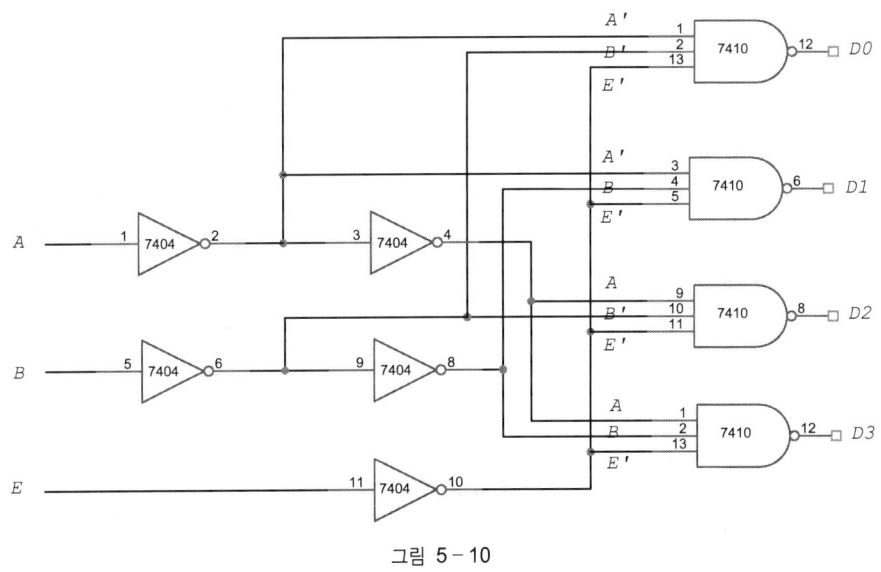

그림 5 - 10

표 5 - 8

E	A	B	D_0	D_1	D_2	D_3
1	x	x				
0	0	0				
0	0	1				
0	1	0				
0	1	1				

■ 논리식

(1) D0 =

(2) D1 =

(3) D2 =

(4) D3 =

(5) 어떤 회로인가?

표 5-6

입력			출력(F)			
S	A	B	D_0	D_1	D_2	D_3
0	0	0				
0	0	+5V				
0	+5V	0				
0	+5V	+5V				
+5V	0	0				
+5V	0	+5V				
+5V	+5V	0				
+5V	+5V	+5V				

■ 논리식

(1) F =

(2) 어떤 회로인가?

표 5-7

입력		출력
A	B	F
0	0	
0	+5V	
+5V	0	
+5V	+5V	

■ 논리식

(1) F =

(2) 어떤 회로인가?

표 5-8

E	A	B	D_0	D_1	D_2	D_3
1	x	x				
0	0	0				
0	0	1				
0	1	0				
0	1	1				

■ 논리식

(1) D0 =

(2) D1 =

(3) D2 =

(4) D3 =

(5) 어떤 회로인가?

승부차기를 대비하라! **인생**에도 누군가가 대신할 수 **없는 순간**이 온다.

실험 6 　　RS 및 D 플립플롭

실험목적

- 기억소자의 기본 개념을 이해할 수 있도록 한다.
- RS 플립플롭의 원리를 이해하고 회로를 설계, 제작할 수 있도록 한다.
- D 플립플롭의 구성방법 및 응용방법을 이해할 수 있도록 한다.
- 채터링 방지 회로의 원리를 이해할 수 있도록 한다.

실험기기 및 재료

구분	품명	규격	수량	비고
기기	논리회로 실험장치		1	
	회로시험기		1	
재료	NAND 게이트	IC 74LS00	2	
	NOR 게이트	IC 74LS02	1	
	컨덴서	0.03μF	1	
	저항	1kΩ	2	
	저항	680Ω	1	
	점퍼선		약간	

어느 디지털 시스템의 출력이 시간에 관계없이 다만 그때의 입력값에 의해서만 결정되는 것을 조합논리(組合論理)라고 하고 그런 회로를 조합회로(組合回路)라고 한다. 반대로 출력이 입력의 조합에 의해서만 결정되는 것이 아니라 선행된 입력에 의해 시간의 관계를 갖는, 즉 후행출력값이 선행출력값에 영향을 받는 것을 순서논리(順序論理)라고 하고 그런 회로를 순서회로(順序回路)라고 한다.

이러한 순서논리를 수행하는 기본 소자를 플립플롭(flip-flop)이라 한다. 또한 플립플롭 회로는 쌍안정 멀티바이브레이터라고도 하고, '1' 및 '0' 두 개의 안정한 상태를 가지며 입력신호에 따라 어느 쪽이 안정 상태를 가지느냐가 결정되는 1비트(One bit)의 정보를 기억할 수 있는 기억회로이다. 일반적으로 입력회로의 구성에 따라서 RS, T, D, JK 플립플롭 등으로 구분된다.

1. RS 플립플롭(RS Flip-Flop)

(1) RS 래치(RS Latch)

RS 플립플롭은 리셋(Reset)과 셋(Set) 입력 단자와 2개의 출력 Q와 $\overline{Q}$로 구성된 비동기형 순서회로를 말한다. 이 회로는 'RS 래치'라고도 하는데 2개의 NAND 게이트나 NOR 게이트를 이용하여 2단으로 정궤환 루프를 구성하면 된다. 이 회로가 그림 6-1과 6-2에 나타나 있다. 이 회로의 동작 과정을 살펴보자.

♪ 그림 6-1의 NAND 게이트로 만들어진 RS FF의 동작 과정을 살펴보면,

R=0, S=0이면 Q_{n+1}과 $\overline{Q}_{n+1} = 1$이 되어 Q_{n+1}과 $\overline{Q}_{n+1}$가 서로 다른 값을 가져야 하는 보수관계를 위반하게 되므로 부정(否定)상태라고 하며 R=0, S=1이

면 $Q_{n+1}=0$, $\overline{Q}_{n+1}=1$이 되고, R=1, S=0이면 $Q_{n+1}=1$, $\overline{Q}_{n+1}=0$이고, R= 1, S=1이면 $Q_{n+1}=Q_n$, $\overline{Q}_{n+1}=\overline{Q}_n$으로 전 상태를 그대로 유지하게 된다. 이 회로의 진리표를 보면 동작 특성을 확실히 알 수 있다. 표 6-1에 NAND 게이트 로 만들어진 RS 플립플롭의 진리표가 있다.

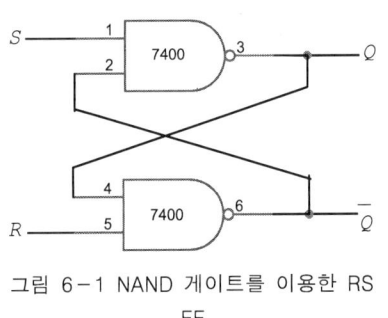

그림 6-1 NAND 게이트를 이용한 RS FF

표 6-1 NAND 게이트 RS FF 진리표

T_n에서의 입력		T_{n+1}에서의 출력		상태
R	S	Q_{n+1}	$\overline{Q}_{n+1}$	
0	0	1	1	부정(사용 않음)
0	1	0	1	Reset
1	0	1	0	Set
1	1	Q_n	$\overline{Q}_n$	원상태 유지

■그림 6-2의 NOR 게이트로 만들어진 RS FF의 동작 과정을 살펴보면,

R=0, S=0 $\Rightarrow Q_{n+1}=Q_n$, $\qquad \overline{Q}_{n+1}=\overline{Q}_n$

R=0, S=1 $\Rightarrow Q_{n+1}=1$, $\qquad \overline{Q}_{n+1}=0$

R=1, S=0 $\Rightarrow Q_{n+1}=0$, $\qquad \overline{Q}_{n+1}=1$

R=1, S=1 $\Rightarrow Q_{n+1}=0$, $\qquad \overline{Q}_{n+1}=0$

이 되어 보수관계를 위반하게 되므로 부정(否定)상태라고 하며 표 6-2에 NOR 게이트로 만들어진 RS 플립플롭의 진리표가 있다.

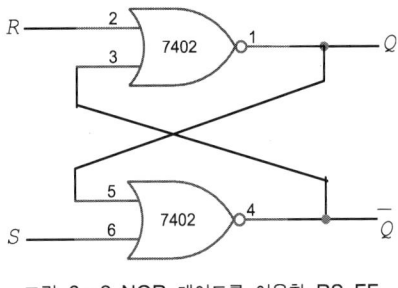

그림 6-2 NOR 게이트를 이용한 RS FF

표 6-2 NOR 게이트 RS FF 진리표

T_n에서의 입력		T_{n+1}에서의 출력		상태
R	S	Q_{n+1}	$\overline{Q}_{n+1}$	
0	0	Q_n	$\overline{Q}_n$	원상태 유지
0	1	1	0	Set
1	0	0	1	Reset
1	1	0	0	부정(사용 않음)

(2) RST 플립플롭(Clocked RS FF)

앞에서 설명한 RS 플립플롭들은 Set, Reset 입력이 언제 유효하게 되는가 하는 시간적 제약이 전혀 없다. 따라서 동기(同期)를 시킬 수가 없으므로 아래 그림과 같이 클럭 입력(C_p)이 '1'일 때만 S=1일 때 Q=1이 되고, C_p가 '0'일 때는 R, S가 어떤 상태라도 $Q_{n+1} = Q_n$으로 상태 변동이 없도록 만든 회로를 말하며 이 회로를 RST 플립플롭 또는 클럭부(附: Clocked) RS 플립플롭이라고 한다. 즉 Clock pulse가 들어올 때 R과 S로 준비된 다음 상태가 된다.

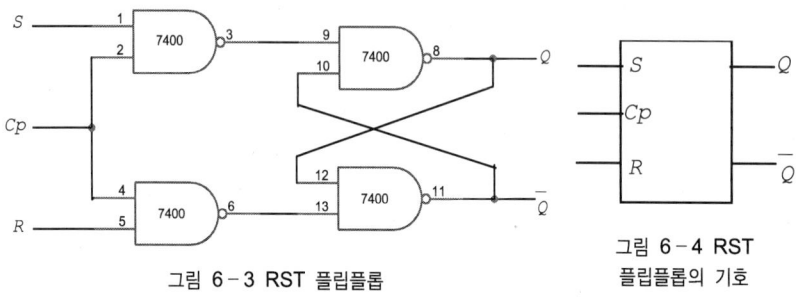

그림 6-3 RST 플립플롭

그림 6-4 RST
플립플롭의 기호

위 회로에서 사용된 클럭펄스(C_p)는 순서회로의 전 플립플롭에 공급되어 모든 플립플롭이 동시에 C_p의 통제를 받아 동작하도록 한다. 이와 같이 시간적 통일성을 부여하기 위해 사용되는 클럭을 마스터 클럭(Master clock)이라 하고 마스터 클럭을 사용한 시스템을 '동기식 시스템'이라 하고 사용하지 않은 시스템을 '비동기식 시스템'이라 한다.

RST 플립플롭의 진리표는 표 6-3과 같다.

표 6-3 진리표

C_p	S	R	Q(t+T)
0	x	x	Q(t)
1	0	0	Q(t)
1	0	1	0
1	1	0	1
1	1	1	부정(사용 않음)

표 6-4 특성표

Q(t)	S	R	Q(t + T)
0	0	0	0
0	0	1	0
0	1	0	1
0	1	1	부정(사용 않음)
1	0	0	1
1	0	1	0
1	1	0	1
1	1	1	부정(사용 않음)

위 특성표에서 RS 플립플롭의 특성방정식은 아래와 같이 유도된다.

RS FF 특성방정식 $Q(t+T) = S + Q(t)R'$

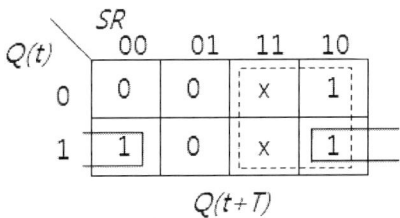

$Q(t+T)$

그림 6-5 RST 플립플롭의 S, R, Cp 입력에 대한 출력 Q의 파형은 아래 그림과 같다(단 이때 플립플롭의 초기 상태가 리셋(Reset) 상태라고 가정함).

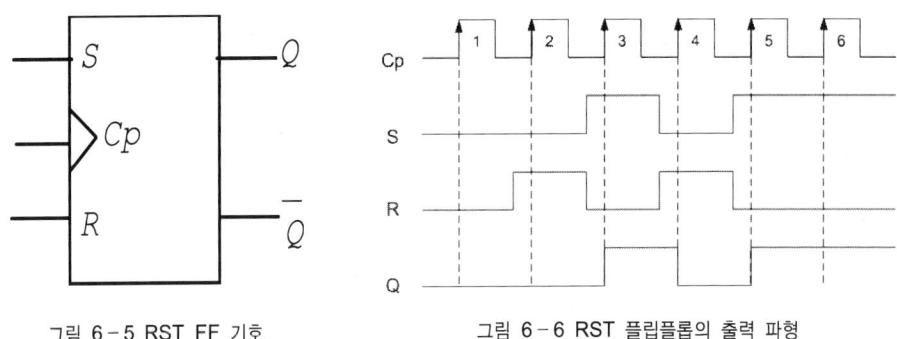

그림 6-5 RST FF 기호   그림 6-6 RST 플립플롭의 출력 파형

위 그림에서

클럭펄스가 '1'일 때 S＝0, R＝0이므로 Q는 변화가 없고

클럭펄스가 '2'일 때 S＝0, R＝1이므로 Q＝0(리셋)

클럭펄스가 '3'일 때 S＝1, R＝0이므로 Q＝1(셋)

클럭펄스가 '4'일 때 S＝0, R＝1이므로 Q＝0(리셋)

클럭펄스가 '5'일 때 S＝1, R＝0이므로 Q＝1(셋)

클럭펄스가 '6'일 때 S＝1, R＝0이므로 Q＝1(셋)이 된다.

(3) SET 및 RESET 우선 RS 플립플롭(Set and Reset priority RS FF)

RS FF에서 금지되어 있는 부정의 입력 상태, 즉 R＝1, S＝1인 경우에도 가능하게 해 주는 회로가 Set 우선 RS 플립플롭이다. 이 회로는 R＝1, S＝1이 입력되는 경우 Set을 우선으로 처리하여 $Q_{n+1}=1, \overline{Q}_{n+1}=0$으로 처리해 준다. 그 회로와 진리표를 그림 6－7과 표 6－5에서 보여준다.

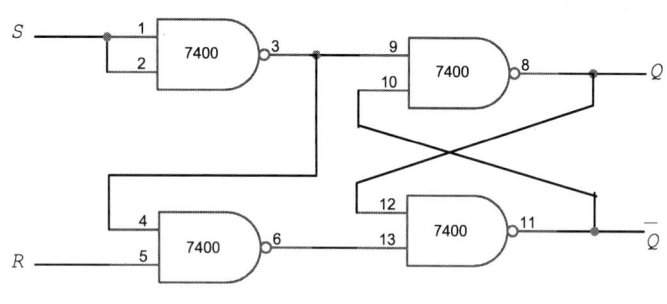

그림 6－7 Set 우선 RS 플립플롭

표 6-5 Set 우선 RS FF 진리표

T_n에서의 입력		T_{n+1}에서의 출력		상태
R	S	Q_{n+1}	$\overline{Q}_{n+1}$	
0	0	Q_n	$\overline{Q}_n$	원상태 유지
0	1	1	0	Set
1	0	0	1	Reset
1	1	1	0	Set 우선

Set 우선 RS 플립플롭에도 클럭을 사용할 수가 있는데 관련 회로가 그림 6-8에 있다. Reset 우선 RS 플립플롭의 동작은 Set 우선 RS 플립플롭과 반대로 생각하면 된다. 리셋 단자에 NOT 회로를 사용하도록 구성하면 이의 동작을 수행토록 할 수 있다.

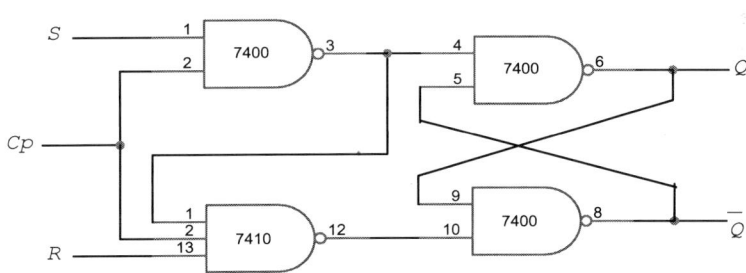

그림 6-8 Set 우선 RST 플립플롭

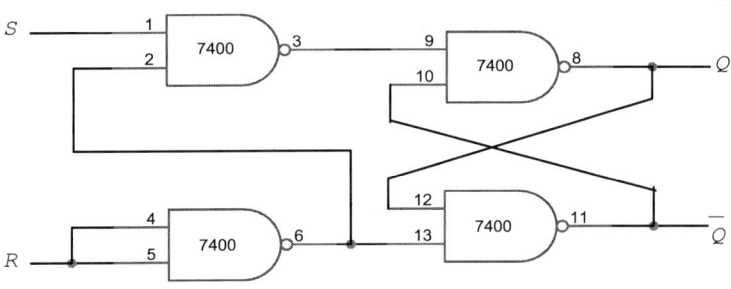

그림 6-9 Reset 우선 RS 플립플롭

(4) 채터링(Chattering) 방지회로

디지털 시스템에서 스위치를 사용할 때 일반적으로 토글(Toggle) 스위치나 계전기 접점을 많이 이용하게 된다. 그런데 이런 기계적인 접점은 전환 시 그림 6-10(a)와 같이 불연속적인 동작을 하게 되는데 이를 채터링이라고 한다. 채터링 현상이 발생하면 여러 개의 펄스가 더 인가되는 것과 같은 현상을 초래하게 되어 순서논리회

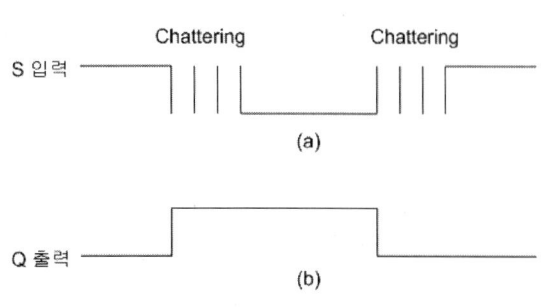

그림 6-10 채터링 입력 및 채터링이 제거된 출력 파형

로에서는 논리의 상태를 전혀 엉뚱하게 바꾸어 놓게 된다. 특히 단일 펄스가 사용되는 시스템에서는 이 회로가 필수적으로 사용된다.

채터링 방지회로는 RS 플립플롭을 이용하여 구성할 수가 있는데 그림 6-11에서 회로를 보여준다. 그림에서 스위치가 R에서 S로 전환 시 S에 일단 접속된 후기계적인 탄성에 의해 잠시 어느 곳에도 접속되어 있지 않은 상태에 있다고 해도 이때는 R=1, S=1이므로 RS 플립플롭의 동작특성에 의해 전 상태를 그대로 유지하게 된다. 따라서 채터링 현상이 방지되고 안정한 동작을 할 수 있게 된다.

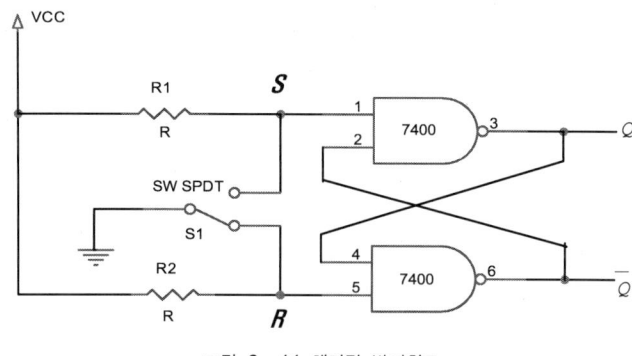

그림 6-11 채터링 방지회로

2. D 플립플롭(D Flip-Flop)

D 플립플롭은 Data 또는 Delayed 플립플롭의 약칭으로 지연형 플립플롭이며 하나의 클럭 입력과 하나의 데이터 입력을 갖는 회로이다. 이는 클럭펄스에 동기되어 있지 않은 입력을 받아서 동기된 출력을 만들어 낸다. 특히 D 플립플롭은 1bit를 저장하기에 매우 유용한 회로로 동기입력을 가진 RS 플립플롭을 변형한 것이다. 특히 이 회로의 특성은 D 입력이 그대로 출력에 전달된다는 것이다. 이 플립플롭은 1bit 타임의 지연소자로 하나의 입력에 의해서 출력 Q가 1bit 타임 전의 상태와 같게 되는 회로이다. 그림 6-10에 NAND 게이트를 이용한 D 플립플롭의 회로를 보여준다.

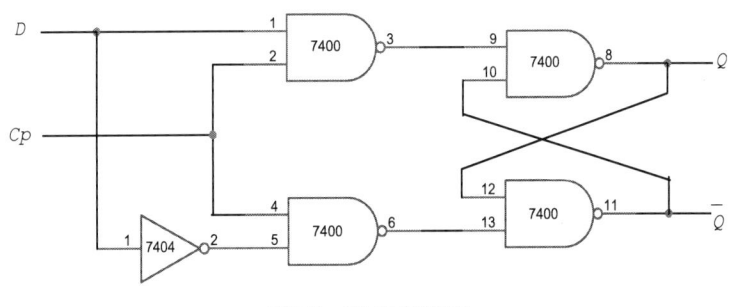

그림 6-12 D 플립플롭

표 6-6 진리표

C_p	D	Q(t+T)
0	x	Q(t)
1	0	0
1	1	1

그림 6-13 D 플립플롭의
기호

표 6-7 진리표

Q_t	D	Q(t+T)
0	0	0
0	1	1
1	0	0
1	1	1

D 플립플롭의 특성방정식은 특성표로부터 구할 수가 있다.

D FF 특성방정식	$Q(t + T) = D$

그림 6 – 14는 D 플립플롭의 출력 Q의 파형이다.
(단 플립플롭이 셋 상태에서 시작한다고 가정함)

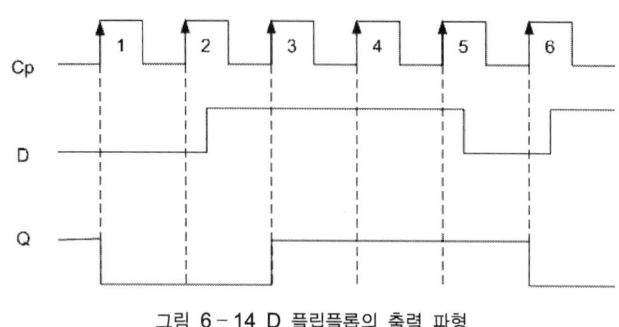

그림 6 – 14 D 플립플롭의 출력 파형

앞의 그림에서 클럭펄스의 리딩 에지(Leading edge)에서 D 입력의 상태가 출력 Q로 나타난다.

클럭펄스가 '1'의 리딩 에지에서 D = 0이므로 Q = 0
클럭펄스가 '2'의 리딩 에지에서 D = 0이므로 Q = 0
클럭펄스가 '3'의 리딩 에지에서 D = 1이므로 Q = 1
클럭펄스가 '4'의 리딩 에지에서 D = 1이므로 Q = 1
클럭펄스가 '5'의 리딩 에지에서 D = 1이므로 Q = 1
클럭펄스가 '6'의 리딩 에지에서 D = 0이므로 Q = 0이 된다.

스타(star)는 팀 속에서 더 빛난다!

(1) 그림 6 − 15와 같이 회로를 구성하고 출력 상태를 측정하여 표 6 − 8에 기록하시오.

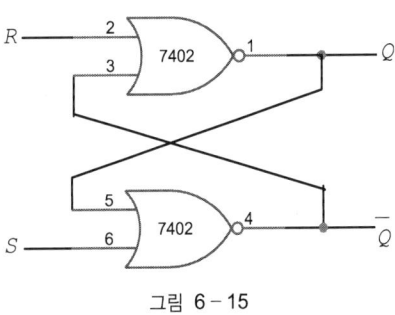

그림 6 − 15

표 6 − 8

R	S	Q	$\overline{Q}$
1	0		
0	0		
0	1		
0	0		

(2) 그림 6 − 16과 같이 회로를 구성하고 출력 상태를 측정하여 표 6 − 9에 기록하시오.

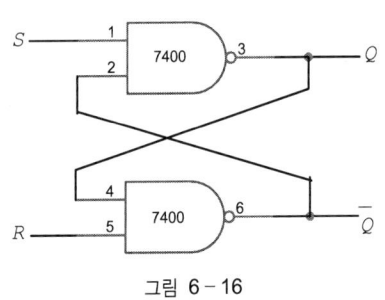

그림 6 − 16

표 6 − 9

R	S	Q	$\overline{Q}$
1	0		
1	1		
0	1		
1	1		

(3) 그림 6 - 17과 같이 회로를 구성하고 출력 상태를 측정하여 표 6 - 10에 기록
   하시오.

   (반드시 표 6 - 10에 기록된 순서대로 스위치를 동작시킬 것)

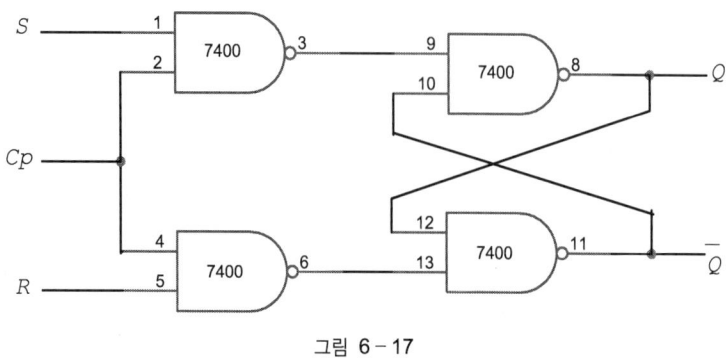

그림 6 - 17

표 6 - 10

순서	Cp	S	R	Q	$\overline{Q}$
1	+5V	0	+5V		
2	0	0	+5V		
3	0	0	0		
4	+5V	0	0		
5	+5V	+5V	0		
6	0	+5V	0		
7	0	0	0		
8	+5V	0	0		
9	+5V	+5V	+5V		

위 그림 6 - 17에서 구성한 회로를 그대로 사용하여 다음 실험을 진행하시오.

(4) 다음의 지시에 의거 실험을 진행하시오.

❶ 그림 6-18과 같이 단일 펄스 발생기를 구성하시오.

단 스위치를 누르지 않은 상태에서 접속 위치는 항상 그림의 위치에 있도록 할 것.

❷ ❶의 회로를 그림 6-17의 Cp 스위치에 연결하고 표 6-11에서 지시하는 순서에 따라 실험하고 측정결과를 기록하시오.

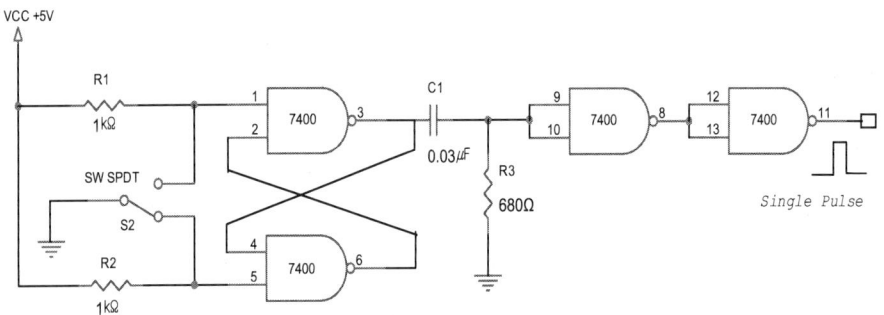

그림 6-18 단일 펄스 발생기

표 6-11

순서	입력		단일 펄스 인가 전의 출력		단일 펄스 인가 후의 출력	
	S	R	Q_n	$\overline{Q_n}$	Q_n	$\overline{Q_n}$
1	+5V	0	- - -	- - -	①	②
2	0	+5V	①	②	③	④
3	+5V	0	③	④	⑤	⑥
4	0	+5V	⑤	⑥	⑦	⑧
5	0	0	⑦	⑧	⑨	⑩
6	+5V	0	⑨	⑩	⑪	⑫
7	0	+5V	⑪	⑫	!! 주 의 !! 단일 펄스를 인가하지 말 것 ※ 숫자가 같은 원문자는 같은 값임	
8	+5V	0				
9	0	+5V				
10	+5V	0				
11	0	+5V				

(5) 그림 6 – 19의 회로를 구성하여 표 6 – 12의 순서에 의해 결과를 측정하고 표에 기록하시오(단 단일 펄스발생기는 그림 6 – 18의 것을 사용할 것).

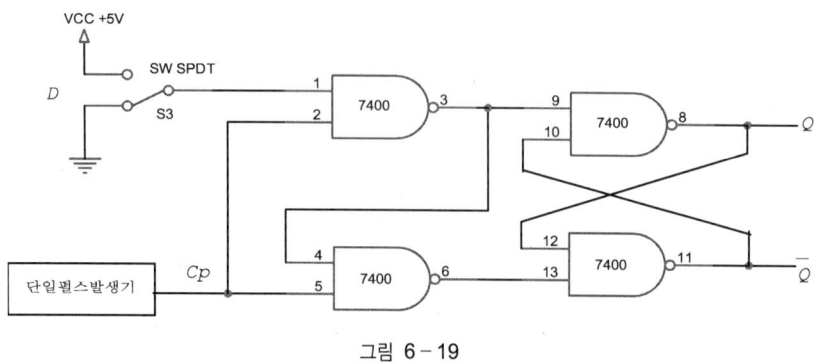

그림 6 – 19

표 6 – 12

순서	입력	단일 펄스 인가 전의 출력		단일 펄스 인가 후의 출력	
	D	Q_n	$\overline{Q}_n$	Q_n	$\overline{Q}_n$
1	+5V	– – –	– – –	①	②
2	0	①	②	③	④
3	+5V	③	④	⑤	⑥
4	0	⑤	⑥	⑦	⑧
5	0	⑦	⑧	⑨	⑩
6	+5V	⑨	⑩	⑪	⑫
7	0	⑪	⑫	!! 주 의 !! 단일 펄스를 인가하지 말 것	
8	+5V				
9	0				
10	+5V				

실험결과 Report		학과명	학번	성명
실험 6	RS 및 D 플립플롭			

표 6-8

R	S	Q	$\overline{Q}$
1	0		
0	0		
0	1		
0	0		

표 6-9

R	S	Q	$\overline{Q}$
1	0		
1	1		
0	1		
1	1		

표 6-10

순서	Cp	S	R	Q	$\overline{Q}$
1	+5V	0	+5V		
2	0	0	+5V		
3	0	0	0		
4	+5V	0	0		
5	+5V	+5V	0		
6	0	+5V	0		
7	0	0	0		
8	+5V	0	0		
9	+5V	+5V	+5V		

표 6-11

순서	입력		단일 펄스 인가 전의 출력		단일 펄스 인가 후의 출력	
	S	R	Q_n	$\overline{Q_n}$	Q_n	$\overline{Q_n}$
1	+5V	0	− − −	− − −	①	②
2	0	+5V	①	②	③	④
3	+5V	0	③	④	⑤	⑥
4	0	+5V	⑤	⑥	⑦	⑧
5	0	0	⑦	⑧	⑨	⑩
6	+5V	0	⑨	⑩	⑪	⑫
7	0	+5V	⑪	⑫		
8	+5V	0			**!! 주 의 !!**	
9	0	+5V			단일 펄스를	
10	+5V	0			인가하지 말 것	
11	0	+5V				

표 6-12

순서	입력	단일 펄스 인가 전의 출력		단일 펄스 인가 후의 출력	
	D	Q_n	$\overline{Q_n}$	Q_n	$\overline{Q_n}$
1	+5V	− − −	− − −	①	②
2	0	①	②	③	④
3	+5V	③	④	⑤	⑥
4	0	⑤	⑥	⑦	⑧
5	0	⑦	⑧	⑨	⑩
6	+5V	⑨	⑩	⑪	⑫
7	0	⑪	⑫		
8	+5V			**!! 주 의 !!**	
9	0			단일 펄스를 인가하지	
10	+5V			말 것	

할 수 있으면 '슛'(Shoot)이다! 할 수 없으면 '패스'(Pass)다! 이 둘 다 중요한 기술이다.

실험 7 JK 및 T 플립플롭

실험목적

- JK 플립플롭의 동작원리를 이해하고 응용할 수 있도록 한다.
- T 플립플롭의 동작원리를 이해하고 회로를 설계, 제작할 수 있도록 한다.
- 플립플롭에서의 레이싱 현상을 이해할 수 있도록 한다.
- 마스터 ─ 슬레이브 플립플롭의 구성을 이해할 수 있도록 한다.

실험기기 및 재료

구분	품명	규격	수량	비고
기기	논리회로 실험장치		1	
	회로시험기		1	
	오실로스코프		1	
	주파수 카운터		1	
재료	NAND 게이트	IC 74LS00	3	2입력
	NAND 게이트	IC 74LS10	2	3입력
	NOR 게이트	IC 74LS02	1	2입력
	NOT 게이트	IC 74LS04	1	
	JK FF	IC 74LS76	1	
	LED		4	
	저항	1kΩ	3	
	저항	330Ω	4	
	점퍼선		약간	

1. JK 플립플롭(JK Flip-Flop)

(1) 궤환형 JK 플립플롭

JK 플립플롭은 RS 플립플롭에서 문제가 되는 불확실한 출력 상태(부정 상태)를 정의하여 사용할 수 있도록 개량된 플립플롭이다. 즉 J, K 두 입력 모두 '1'이 되더라도 불확실한 출력 상태를 갖지 않고 J=K=1인 상태에서 클럭펄스가 인가될 때마다 출력값이 반전되도록 하여 안정한 상태로 동작할 수 있도록 만든 회로이다. 따라서 이 플립플롭은 RS 플립플롭을 이용하여 구성할 수 있는데 그림 7-1에서 그 회로를 보여주고 있다.

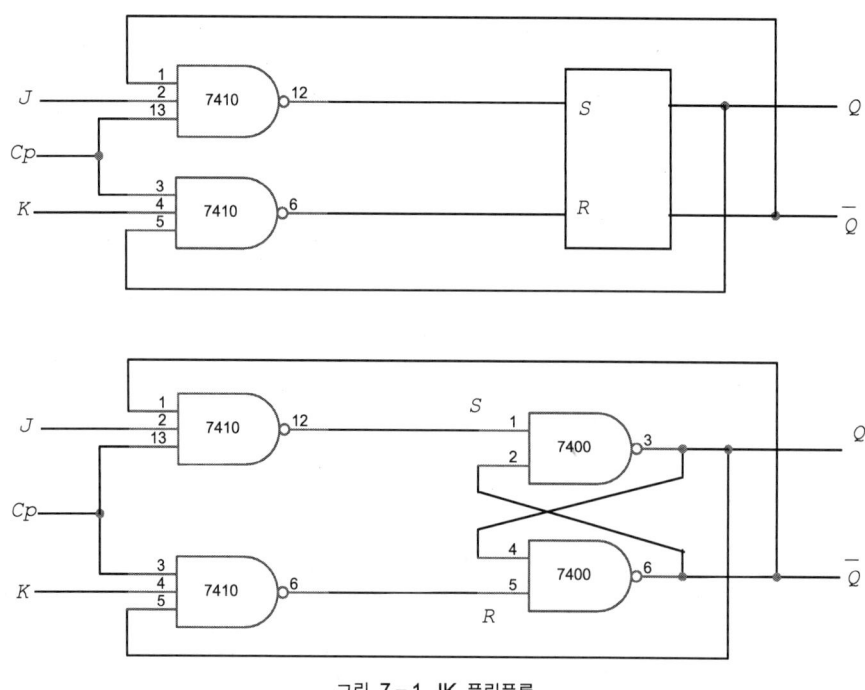

그림 7-1 JK 플립플롭

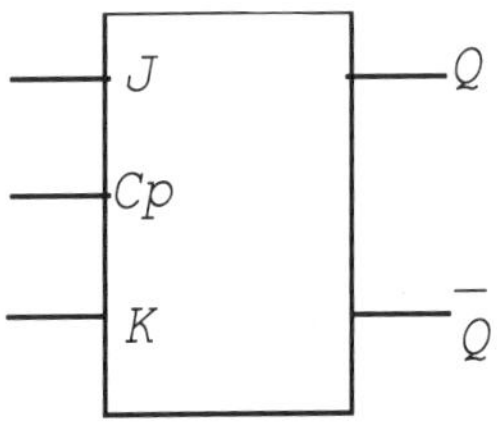

그림 7-2 JK 플립플롭의
기호

표 7-1 진리표

C_p	J	K	Q(t+T)	상태
0	x	x	Q	상태변화 없음
1	0	0	Q	원상태 유지
1	0	1	0	리셋
1	1	0	1	셋
1	1	1	$\overline{Q}$	원상태 반전

표 7-2 특성표

Q(t)	J	K	Q(t+T)
0	0	0	0
0	0	1	0
0	1	0	1
0	1	1	1
1	0	0	1
1	0	1	0
1	1	0	1
1	1	1	0

위 특성표에 의해 JK 플립플롭의 특성방정식을 구하면 아래와 같다.

JK FF 특성방정식	$Q(t+T) = JQ'(t) + K'Q(t)$

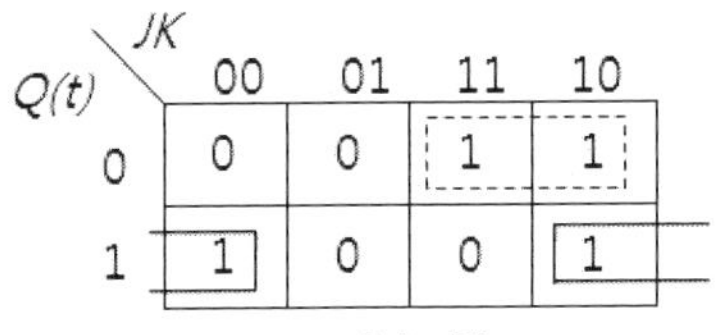

JK 플립플롭에서는 실제로 클럭펄스가 들어오기 전에 초기의 값을 결정하기 위해 그림 7−3과 같이 PR(Peset) 및 CLR(Clear) 단자를 두고 PR=0일 경우 Q=1로 Set 시키고 CLR=0이면 Q=0으로 Reset 시키도록 하여 사용되고 있다.

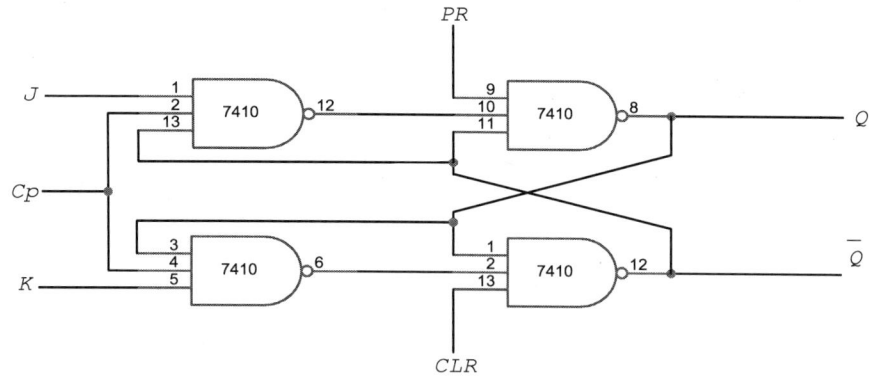

그림 7−3 PR과 CLR 단자가 있는 JK 플립플롭

그림 7−3 JK 플립플롭의 출력 Q의 파형은 아래 그림과 같다.
(단 이때 플립플롭의 초기 상태가 Q=0인 리셋(Reset) 상태라고 가정)

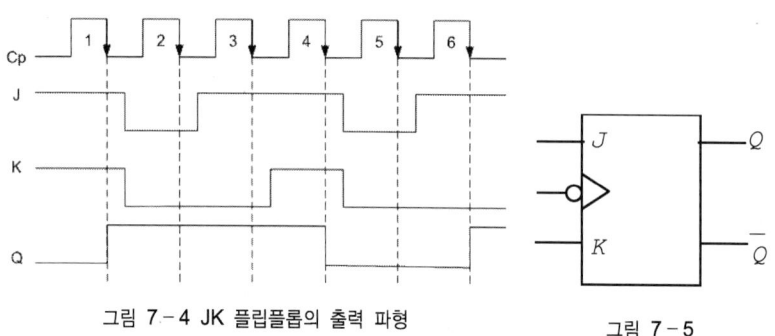

그림 7−4 JK 플립플롭의 출력 파형   그림 7−5

앞 그림에서

클럭펄스가 '1'일 때 J = 1, K = 1이므로 Q = 1로 바뀌고

클럭펄스가 '2'일 때 J = 0, K = 0이므로 Q = 1로 불변

클럭펄스가 '3'일 때 J = 1, K = 0이므로 Q = 1로 셋 상태

클럭펄스가 '4'일 때 J = 1, K = 1이므로 Q = 0로 바뀌고

클럭펄스가 '5'일 때 J = 0, K = 0이므로 Q = 0로 불변

클럭펄스가 '6'일 때 J = 1, K = 0이므로 Q = 1로 셋 상태로 바뀌게 된다.

(2) 레이싱(Racing) 현상

입력클럭펄스는 다음과 같은 두 가지 형태의 펄스를 사용할 수 있다.

클럭펄스의 입력에는 그림과 같이 클럭입력의 논리 상태가 'L'에서 'H'로 변할 때 입력을 읽어 넣으면 동시에 출력이 나타나는 포지티브 에지(Positive edge) 트리거 방식과 클럭 입력이 'H'에서 'L'로 변할 때 출력이 나타나는 네가티브 에지(Negative edge) 트리거 방식이 있다.

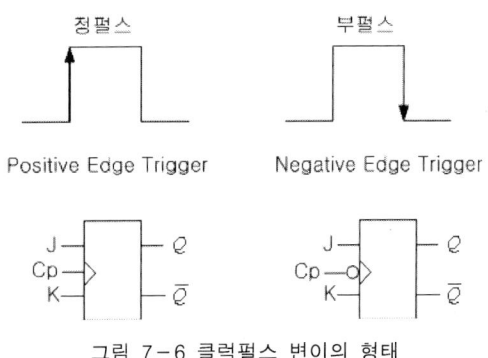

그림 7-6 클럭펄스 변이의 형태

포지티브 에지 트리거로 사용되는 펄스를 정(正)펄스라고 하고 이런 펄스를 사용하는 논리회로를 '정논리회로'라고 한다. 또한 네가티브 에지 트리거로 사용되는 펄스를 부(負)펄스라고 하고 이런 펄스를 사용하는 논리회로를 '부논리회로'라고 한다.

그런데 플립플롭에서 포지티브 에지 트리거를 사용하는 방식은 다음과 같은 문제점이 있다. 즉 JK 플립플롭의 경우 J＝1, K＝1이고 출력 Q가 '0'일 때 클럭펄스 '1'이 가해지면 플립플롭 회로를 전파하는 시간 t만큼 지연된 후 출력 Q＝1이 나타난다. 이때 <u>클럭펄스의 지연 시간이 플립플롭의 지연 시간(t)보다 클 경우 클럭펄스 한 개 입력에 플립플롭의 출력 상태가 여러 번 '0'과 '1'을 반복하여 바뀌게 되어 출력 상태가 불안정하게 된다.</u> 이와 같이 '클럭펄스가 '1'일 때 출력 상태가 변화되면 입력 측에 변화를 일으켜 오동작이 발생되는 현상'을 '**레이싱(Racing) 현상**'이라고 한다.

이 레이싱 현상을 방지하려면

❶ 클럭펄스의 지연 시간을 플립플롭의 지연 시간(t)보다 짧게 하거나

❷ 네가티브 에지 트리거(Negative Edge Trigger) 방식을 사용하거나

❸ 마스터 슬레이브(Master slave) 플립플롭을 사용하면 된다.

여기서는 마스터 슬레이브 플립플롭을 사용하는 방법에 대해 설명하도록 하겠다.

(3) 마스터 슬레이브(Master slave) JK 플립플롭

마스터 슬레이브 플립플롭은 그림 7 – 7과 같이 마스터 플립플롭과 슬레이브 플립플롭으로 구성되어 있는 일종의 RST 플립플롭 2단에 의한 시프트레지시터(Shift register)로 마스터 측에는 C_p를 슬레이브 측에는 $\overline{C_p}$를 가하고 있다. Q와 $\overline{Q}$ 출력은 $C_p = 1$일 때 마스터 측에 궤환되어 첫 번째 NAND 게이트의 상태를 결정하나 이때 $\overline{C_p} = 0$이므로 세 번째의 NAND 게이트 출력으로는 전달될 수 없고 Q와 $\overline{Q}$의 상태는 변하지 않는다. $C_p = 0$이 되면 Q와 $\overline{Q}$의 상태가 첫 번째 NAND 게이트에 전달되는 것은 저지하나($C_p = 0$이면 첫 번째 NAND 게이트의 출력은 모두 '1') $\overline{C_p} = 1$이므로 마스터 측의 출력이 슬레이브 측으로 전달되어 Q와 $\overline{Q}$의 상태를 역전시킨다. 따라서 이 플립플롭의 진리표는 표 7 – 3과 같이 나타낼 수가 있다. 또한 J＝K＝1일 때 C_p에 의한 NAND 게이트의 변화 상태는 표 7 – 4와 같이 나타낼 수 있다. 결국 이 마스터 – 슬레

이브 JK 플립플롭은 금지입력이 없고 C_p의 펄스폭에 따른 레이싱 현상이 없으며 펄스의 상승 시간 여하에 따른 오동작도 방지할 수 있게 된다.

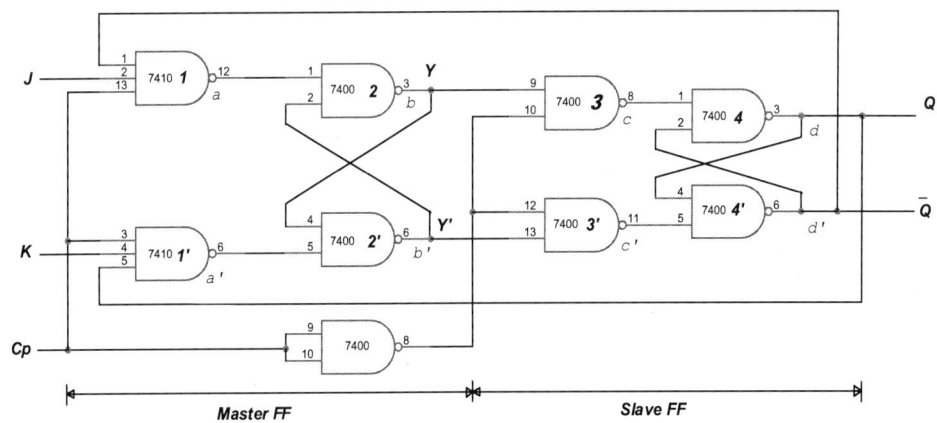

그림 7 − 7 마스터 − 슬레이브 JK 플립플롭

표 7 − 3 마스터 − 슬레이브 JK FF 진리표

T_n에서의 입력		클럭펄스	T_{n+1}에서의 출력	
J	K		Q_{n+1}	$\overline{Q}_{n+1}$
0	0	X	Q_n	$\overline{Q}_n$
0	1	0→1→0	0	1
1	0	0→1→0	1	0
1	1	0→1→0	$\overline{Q}_n$	Q_n

표 7 - 4 클럭펄스에 의한 상태 변화표

J＝K＝1	NAND1		NAND2		NAND3		NAND4	
Cp	a	$\overline{a}$	b	$\overline{b}$	c	$\overline{c}$	d	$\overline{d}$
0	1	1	0	1	1	0	0	1
1	0	1	1	0	1	1	0	1
0	1	1	1	0	0	1	1	0
1	1	0	0	1	1	1	1	0
0	1	1	0	1	1	0	0	1
1	0	1	1	0	1	1	0	1

2. T 플립플롭(T Flip - Flop)

　T 플립플롭은 트리거링(Triggering) 또는 토글링(Toggling) 플립플롭이라고도 하며 T 입력 1개와 Q 및 $\overline{Q}$ 출력 2개를 갖는 회로로 입력 단자에 클럭이 있을 때마다 출력이 바뀌는 플립플롭으로 JK 플립플롭에서 J＝K＝1의 상태가 된다. 아래 그림은 RST, D, JK 플립플롭으로 만들 수 있는 T 플립플롭의 각종 형태를 보여준다.

　이 플립플롭은 별도로 IC화되어 있지는 않다. 따라서 이 형태를 잘 익혀 두기 바란다.

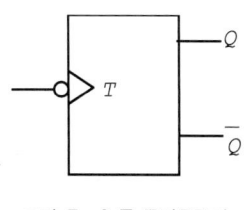

그림 7 - 8 T 플립플롭의 기호

표 7 - 5 진리표

T	Q(t+T)
0	$Q(t)$
1	$\overline{Q(t)}$

표 7 - 6 특성표

Q(t)	T	Q(t+T)
0	0	0
0	1	1
1	0	1
1	1	0

특성표에서 T=1일 경우는 출력값이 전 값의 반대 값으로 바뀌게 되므로 T FF 의 특성방정식은 아래와 같이 된다.

| T FF 특성방정식 | $Q(t+T) = TQ'(t) + T'Q(t) = T \oplus Q(t)$ |

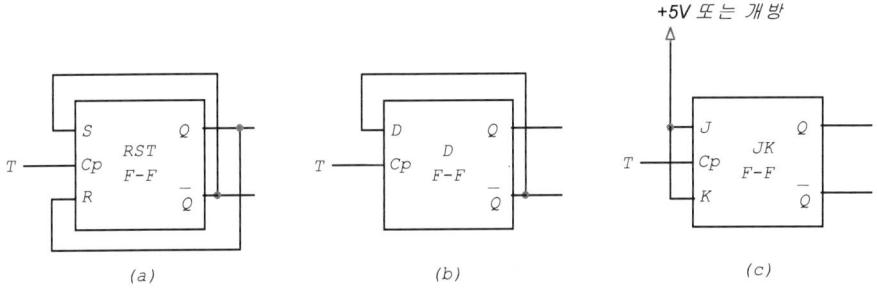

그림 7-9 RST, D, JK 플립플롭을 이용한 T 플립플롭의 구성

T 플립플롭의 입력에 클럭펄스를 인가 시 출력 Q의 파형은 아래 그림과 같다. 앞 그림에서 출력값은 클럭의 하강 모서리에서 변화하므로

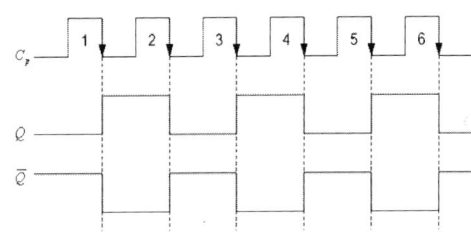

그림 7-10 T 플립플롭의 출력 파형

클럭펄스가 '1'일 때 클럭입력 T=1이므로 Q=0→1로

클럭펄스가 '2'일 때 클럭입력 T=1이므로 Q=1→0으로

클럭펄스가 '3'일 때 클럭입력 T=1이므로 Q=0→1로

클럭펄스가 '4'일 때 클럭입력 T=1이므로 Q=1→0으로

클럭펄스가 '5'일 때 클럭입력 T=1이므로 Q=0→1로 바뀌게 된다.

앞의 그림에서도 알 수 있듯이 T형 플립플롭의 1단 출력은 입력 주파수의 1/2이므로 n개의 T형 플립플롭을 직렬로 연결하면 최종단의 출력 주파수는 1/2n이 된다. 이 특성을 이용하여 T형 플립플롭은 주파수 분주회로에 사용되기도 한다.

참고로 그림 7 – 9(c)에 있는 JK 플립플롭을 이용하여 T 플립플롭을 구성한 회로를 상세히 그리면 아래 그림과 같다.

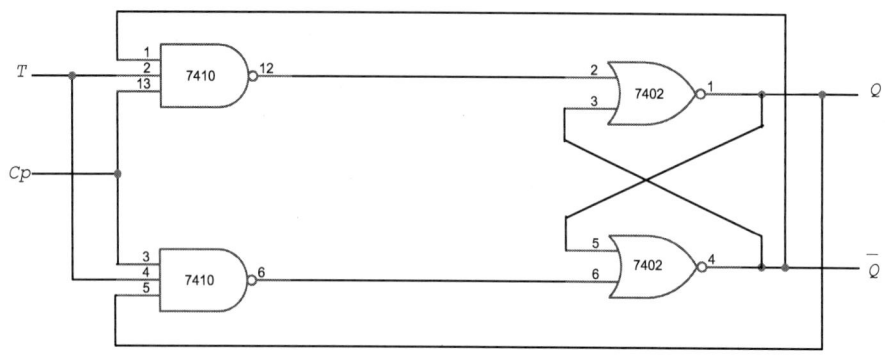

그림 7 – 11 JK 플립플롭을 이용한 T형 플립플롭 구성의 상세도

(1) 그림 7 - 12와 7 - 13의 회로를 구성하고 출력 상태를 측정하여 표 7 - 7과 7 - 8에 기록한 후 결과를 서로 비교하시오.

☝ 유의사항

❶ 반드시 표에서 지시하는 실험 순서를 지킬 것

❷ 초기 상태의 선정방법은 $C_P = 0$으로 한 후

◇ 리셋 상태의 경우: CLR 단자를 잠시 접지('0')시킨 후 개방 상태로 할 것

◇ 셋 상태의 경우: PR 단자를 잠시 접지('0')시킨 후 개방 상태로 할 것

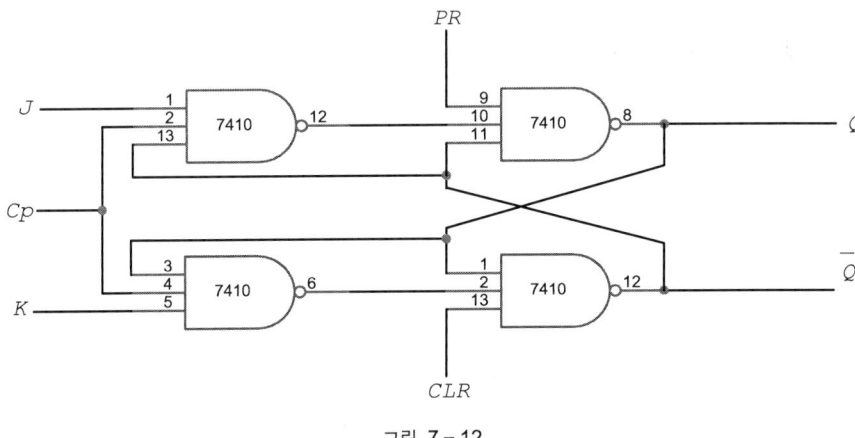

그림 7 - 12

☝ 주의!!!

4→5, 10→11로 변화시킬 때는 반드시 J를 먼저 +5V에서 0으로 한 후 다음에 K를 0에서 +5V로 할 것

표 7-7

초기 상태	순서	Cp	J	K	Q	$\overline{Q}$
리셋 (Reset)	1	0	0	+5V		
	2	0	+5V	0		
	3	+5V	0	0		
	4	+5V	+5V	0		
	5	+5V	0	+5V		
	6	+5V	+5V	+5V		
셋 (Set)	7	0	0	+5V		
	8	0	+5V	0		
	9	+5V	0	0		
	10	+5V	+5V	0		
	11	+5V	0	+5V		
	12	+5V	+5V	+5V		

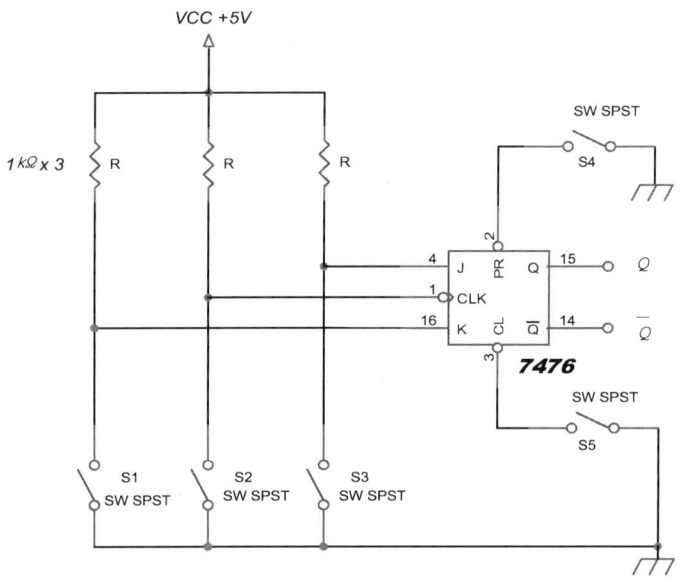

그림 7 - 13

표 7 - 8

초기 상태	순서	Cp	J	K	Q	$\overline{Q}$
리셋 (Reset)	1	0	0	+5V		
	2	0	+5V	0		
	3	+5V	0	0		
	4	+5V	+5V	0		
	5	+5V	0	+5V		
	6	+5V	+5V	+5V		
셋 (Set)	7	0	0	+5V		
	8	0	+5V	0		
	9	+5V	0	0		
	10	+5V	+5V	0		
	11	+5V	0	+5V		
	12	+5V	+5V	+5V		

(2) 그림 7 - 14의 회로를 구성하고 출력 상태를 측정하여 표 7 - 9에 기록하시오.

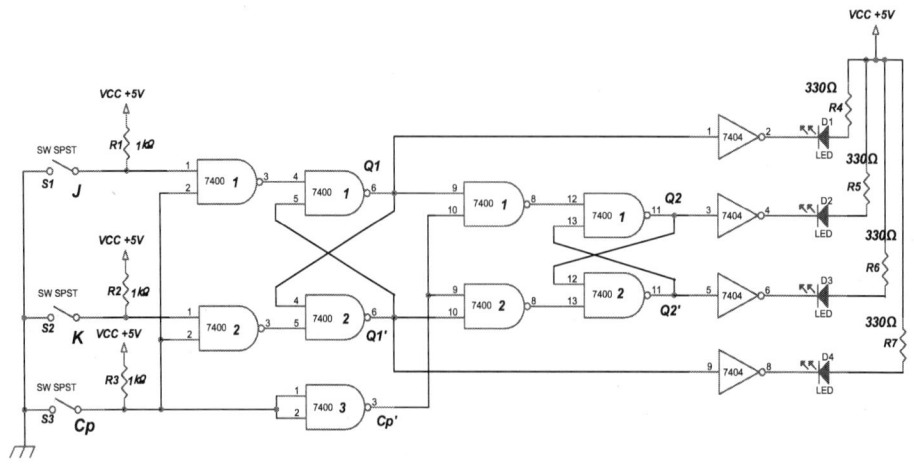

그림 7 - 14 JK 마스터 - 슬레이브 플립플롭

표 7 - 9 진리표

입력			출력			
C_p	K	J	Q_1	$\overline{Q_1}$	Q_2	$\overline{Q_2}$
0	0	0				
0	0	1				
0	1	0				
0	1	1				
1	0	0				
1	0	1				
1	1	0				
1	1	1				

(3) 그림 7 – 15의 회로를 구성하고 출력 상태를 측정하여 표 7 – 10에 기록하시오 (단 초기 Q의 값을 '0'으로 하고 실험할 것).

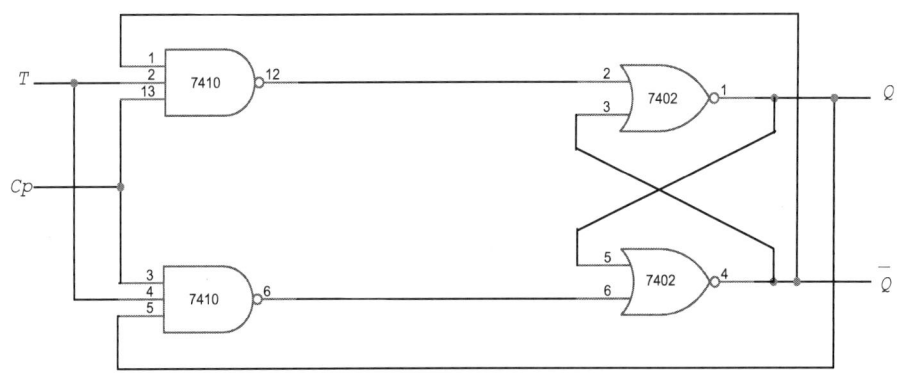

그림 7 – 15

표 7 – 10 진리표

C_p	T	Q(t + T)
1	1	
2	1	
3	1	
4	1	
5	1	
6	1	
7	1	
8	0	
9	0	
10	1	

표 7-7

초기 상태	순서	Cp	J	K	Q	$\overline{Q}$
리셋 (Reset)	1	0	0	+5V		
	2	0	+5V	0		
	3	+5V	0	0		
	4	+5V	+5V	0		
	5	+5V	0	+5V		
	6	+5V	+5V	+5V		
셋 (Set)	7	0	0	+5V		
	8	0	+5V	0		
	9	+5V	0	0		
	10	+5V	+5V	0		
	11	+5V	0	+5V		
	12	+5V	+5V	+5V		

표 7-8

초기 상태	순서	Cp	J	K	Q	$\overline{Q}$
리셋 (Reset)	1	0	0	+5V		
	2	0	+5V	0		
	3	+5V	0	0		
	4	+5V	+5V	0		
	5	+5V	0	+5V		
	6	+5V	+5V	+5V		
셋 (Set)	7	0	0	+5V		
	8	0	+5V	0		
	9	+5V	0	0		
	10	+5V	+5V	0		
	11	+5V	0	+5V		
	12	+5V	+5V	+5V		

표 7-9 진리표

입력			출력			
C_p	K	J	Q_1	$\overline{Q_1}$	Q_2	$\overline{Q_2}$
0	0	0				
0	0	1				
0	1	0				
0	1	1				
1	0	0				
1	0	1				
1	1	0				
1	1	1				

표 7-10 진리표

C_p	T	Q(t+T)
1	1	
2	1	
3	1	
4	1	
5	1	
6	1	
7	1	
8	0	
9	0	
10	1	

어깨에서는 **강속구!** 머리에서는 **유인구!** 인생에서는 **힘**과 **지혜**가 필요하다.

제2부   카운터와 레지스터

실험 8 비동기식 카운터

실험목적

- 카운터의 원리를 이해하고 비동기식 카운터를 통하여 플립플롭의 응용방법을 익히도록 한다.
- 비동기식 카운터를 이용, BCD(10진) 리플 카운터를 설계할 수 있도록 한다.
- 카운터의 응용능력을 배양한다.

실험기기 및 재료

구분	품명	규격	수량	비고
기기	논리회로 실험장치		1	
	회로시험기		1	
	오실로스코프		1	
	주파수 카운터		1	
재료	NAND 게이트	IC 74LS00	1	2입력
	JK FF	IC 74LS73	2	
	점퍼선		약간	

1. 비동기식 카운터(Asynchronous Counter)

(1) 카운터 개요

카운터(Counter)는 데이터를 저장하고 입력된 클럭의 수를 세는 기능이 있다. 즉 1비트의 데이터를 저장하는 메모리 기능을 가지고 있는 플립플롭을 사용하여 입력된 클럭펄스의 수를 기억하도록 한 것이다.

다음 그림을 보자.

아래 그림은 입력클럭이 발생했을 때 변화하는 플립플롭의 출력 상태를 나타낸 것으로 상태천이도라고 한다. 이 상태천이도를 보면 각 원에 있는 숫자는 하나의 가능한 출력 상태(Q_2, Q_1, Q_0)를 나타내는데 이 상태들이 계속하여 000→001→010→011→100→101→110→111→000이 순차적으로 반복해서 나타나고 있는 것을 알 수 있다. 카운터란 바로 이러한 상태를 만들어 주는 회로이다.

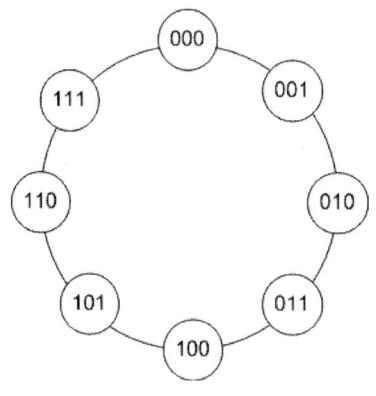

그림 8 - 1 상태천이도

카운터가 카운트할 수 있는 플립플롭 수에 따라 달라진다. 플립플롭의 수가 n개이면 2n개의 수를 카운트할 수 있다.

즉 플립플롭을

❶ 2개 사용할 때 카운트 수: $2^2 = 4$개($0 \rightarrow 1 \rightarrow 2 \rightarrow 3 \rightarrow 0 \cdots\cdots$)

❷ 3개 사용할 때 카운트 수: $2^3 = 8$개($0 \rightarrow 1 \rightarrow 2 \rightarrow 3 \rightarrow 4 \rightarrow 5 \rightarrow 6 \rightarrow 7 \rightarrow 0 \cdots\cdots$)

❸ 4개 사용할 때 카운트 수: $2^4 = 16$개($0 \rightarrow 1 \rightarrow 2 \rightarrow 3 \rightarrow 4 \rightarrow 5 \rightarrow 6 \rightarrow 7 \rightarrow 8 \rightarrow 9 \rightarrow A \rightarrow$ $B \rightarrow C \rightarrow D \rightarrow E \rightarrow F \rightarrow 0 \cdots\cdots$)

를 카운트할 수 있다.

이것을 '**카운팅 결과 나타나는 논리 상태의 수**'라고 정의할 수 있는 'MOD(Modulus) 수'로 표현하면 다음과 같다.

$$\text{MOD 수} = 2^n$$

여기서 카운팅할 수 있는 경우의 수가 N개이면 그 카운터를 **MOD – N 카운터**라고 부른다.

카운터는 **비동기식**(Asynchronous)과 **동기식**(Synchronous) 2가지로 구분할 수 있는데, **비동기식 카운터**는 클럭펄스가 카운터 내의 각 플립플롭의 클럭 단자로 직접 연결되지 않고 앞 단의 플립플롭의 출력이 다음 단 플립플롭의 클럭 단자로 연결되기 때문에 클럭펄스의 입력에 의해 입, 출력 변화가 동시에 일어나지 않고 입력의 변화보다 플립플롭의 동작지연 시간만큼 지연되어서 나타난다. 이러한 카운터를 **리플**(Ripple) **카운터**라고도 한다.

동기식 카운터는 입력클럭펄스가 각 단의 클럭값을 동시에 동기시키는 것으로 **병렬 카운터**라고도 한다. 동기식 카운터는 순차회로에 의한 설계를 할 수 있다. 이번 장에서는 비동기식 카운터에 대해서 다루도록 하겠다.

(2) MOD-4 리플 카운터

MOD-4 리플 카운터는 그림 8-2와 같이 구성할 수 있다. 카운터는 토글 (Toggle)의 원리를 이용하므로 JK 플립플롭 2개를 T형 플립플롭으로 변형시켜 직렬로 접속하여 사용한다.

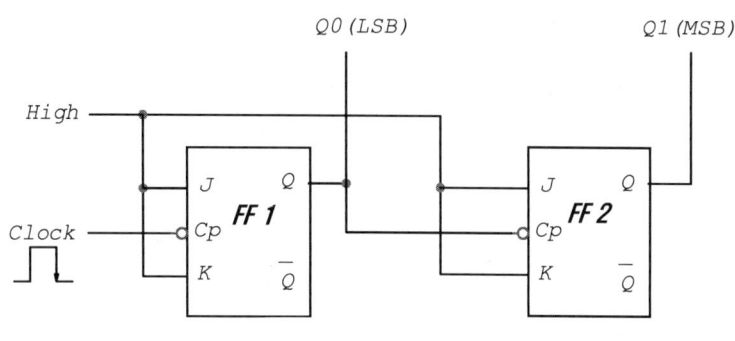

그림 8-2 MOD-4 리플 카운터

위의 그림에서 보면 플립플롭 1(FF 1)의 클럭 입력 단자에는 클럭펄스가 직접 연결되고 플립플롭 2(FF 2)의 클럭 입력 단자는 FF 1의 출력 Q 단자에 연결되어 있음을 알 수 있다. 이 회로는 플립플롭을 2개 사용하므로 출력은 2비트(Q0, Q1)를 가지며 출력 상태는 4가지가 된다. 따라서 이러한 카운터를 MOD-4 리플 카운터라고 한다.

표 8-1은 이 회로의 상태천이표를, 그림 8-3은 카운터 동작의 타이밍도를 나타내는데 4번째 클럭펄스 입력 순간에 FF 2의 출력 Q1이 원래의 상태로 돌아감을 알 수 있다.

표 8-1 MOD-4 리플 카운터 상태천이표

클럭	출력		
(CLK)	Q₁	Q₀	십진수
0	0	0	0
1	0	1	1
2	1	0	2
3	1	1	3
4	0	0	0
5	0	1	1
6	1	0	2
7	1	1	3
8	0	0	0
9	0	1	1
10	1	1	2

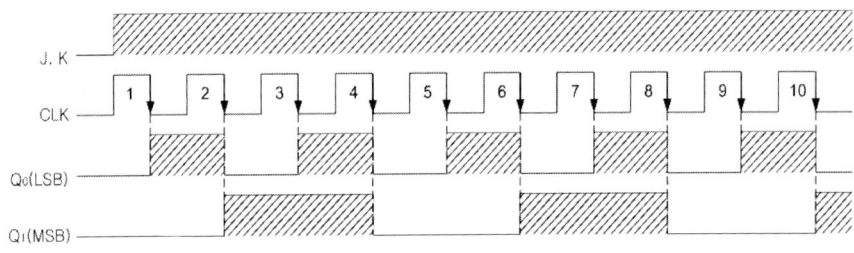

그림 8-3 MOD-4 리플 카운터 타이밍도

위의 개념을 확장하여 MOD-16 리플 카운터에 대해서 살펴보기로 하자.

(3) MOD-16 리플 카운터

MOD-16 리플 카운터는 플립플롭 4개를 직렬로 접속하여 구성한다. 플립플롭 4개를 사용하므로 출력 단자는 4개를 가지며 MOD 수는 16이 되어 MOD-16 리플 카운터가 된다. 이 카운터는 16번째의 클럭펄스가 입력될 때 원래의 상태로 돌아가게 된다.

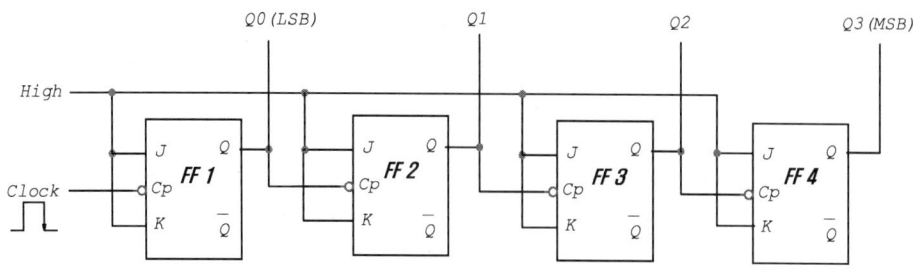

그림 8-4 MOD-16 리플 카운터

표 8-2 MOD-16 리플 카운터 상태천이표

클럭 (CLK)	출력					클럭 (CLK)	출력				
	Q_3	Q_2	Q_1	Q_0	십진수		Q_3	Q_2	Q_1	Q_0	십진수
0	0	0	0	0	0	16	0	0	0	0	0
1	0	0	0	1	1	17	0	0	0	1	1
2	0	0	1	0	2	18	0	0	1	0	2
3	0	0	1	1	3	19	0	0	1	1	3
4	0	1	0	0	4	20	0	1	0	0	4
5	0	1	0	1	5	21	0	1	0	1	5
6	0	1	1	0	6	22	0	1	1	0	6
7	0	1	1	1	7	23	0	1	1	1	7
8	1	0	0	0	8	24	1	0	0	0	8
9	1	0	0	1	9	25	1	0	0	1	9
10	1	0	1	0	10	26	1	0	1	0	10
11	1	0	1	1	11	27	1	0	1	1	11
12	1	1	0	0	12	28	1	1	0	0	12
13	1	1	0	1	13	29	1	1	0	1	13
14	1	1	1	0	14	30	1	1	1	0	14
15	1	1	1	1	15	31	1	1	1	1	15

그림 8-4는 앞 단 플립플롭의 출력 Q가 후단 플립플롭 클럭 단자로 인가되고 있으며, 표 8-2에서 카운트 상태가 0→1→2→3→4→……식으로 증가하는 방향으로 변화하고 있는 것을 알 수 있다. 이러한 카운터를 **비동기식 가산**(加算) **카운터**라고 하며, 반대로 카운트 상태가 15→14→13→12→11→……식으로 감소하는 방향으로 변화하는 카운터를 간단히 설계할 수가 있는데 그것은 앞 단 플립플롭의 출력 Q를 후단 플립플롭 클럭 단자로 인가하지 않고 앞 단 플립플롭의 출력 $\overline{Q}$를 후단의 클럭 단자로 인가하면 되기 때문이다. 이러한 회로를 **비동기식 감산**(減算) **카운터**라고 한다.

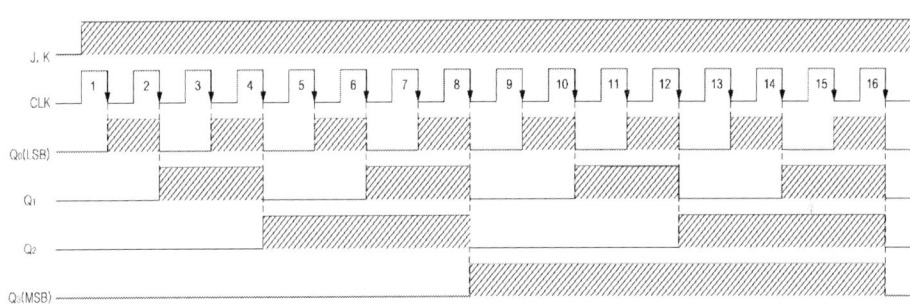

그림 8-5 MOD-16 리플 카운터 타이밍도

의 그림을 보면 플립플롭의 각 단이 모두 2진수의 카운터 동작을 하여 전체적으로 $2^4 = 16$진 카운터로서 동작을 하고 있다는 사실을 알 수 있다. 또한 플립플롭 각 단의 출력은 앞 단의 상태가 '1'에서 '0'으로 바뀔 때마다 변하며 각 단의 주파수는 전단 주파수의 1/2이 된다는 사실도 알 수 있다. 따라서 이러한 카운터는 분주기(分周器)로 사용되기도 한다.

(4) BCD 리플 카운터

BCD 카운터는 2진화 10진 카운터로서 MOD 수가 0~9(0000~1001)까지 변하는 MOD - 10 카운터라고도 한다. 따라서 이 카운터는 플립플롭이 4개가 필요한데 이렇게 되면 앞에서 설명한 MOD - 16 카운터가 되기 때문에 십진수 10에서 15(1010~1111)까지는 사용할 수 없도록 하고 10(1010)이 되면 다시 0(0000)의 상태로 돌아가도록 리셋이 되어야 한다. 이것은 그림 8-5에서 알 수 있듯이 10(1010)의 순간에는 Q3 = Q1 = High 상태이므로 이를 검출하여 그림 8-4의 리셋 (Reset) 입력으로 인가해 주면 된다. 이러한 BCD 리플 카운터를 아래 그림 8-6에서 볼 수 있다.

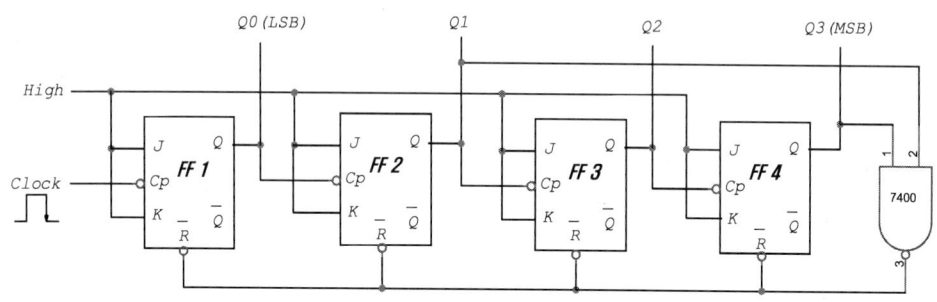

그림 8-6 BCD 리플 카운터

표 8-3 BCD 리플 카운터 상태천이표

클럭 (CLK)	출력					클럭 (CLK)	출력				
	Q_3	Q_2	Q_1	Q_0	십진수		Q_3	Q_2	Q_1	Q_0	십진수
0	0	0	0	0	0	10	0	0	0	0	0
1	0	0	0	1	1	11	0	0	0	1	1
2	0	0	1	0	2	12	0	0	1	0	2
3	0	0	1	1	3	13	0	0	1	1	3
4	0	1	0	0	4	14	0	1	0	0	4
5	0	1	0	1	5	15	0	1	0	1	5
6	0	1	1	0	6	16	0	1	1	0	6
7	0	1	1	1	7	17	0	1	1	1	7
8	1	0	0	0	8	18	1	0	0	0	8
9	1	0	0	1	9	19	1	0	0	1	9

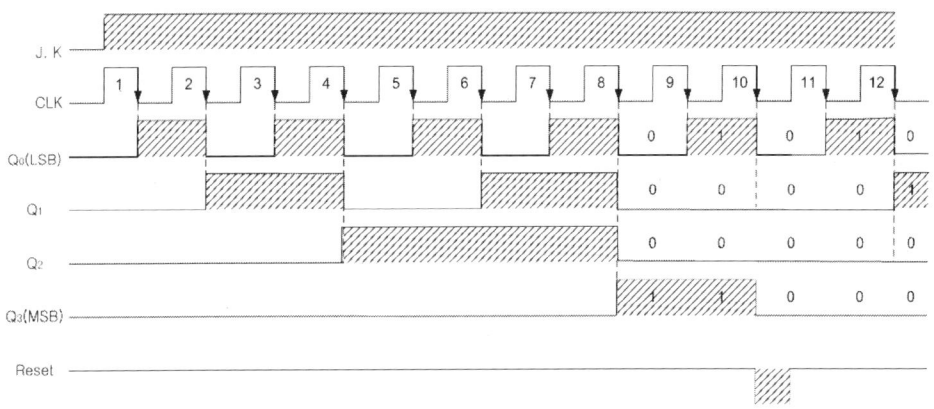

그림 8 - 7 BCD 리플 카운터 타이밍도

(5) 리플 카운터의 전파지연

앞에서 설명한 바와 같이 리플 카운터는 각 단의 플립플롭이 한 번의 클럭 입력에 동시에 동작하지 못하고 앞 단 플립플롭의 출력이 후단 플립플롭의 클럭 단자로 입력되는 직렬접속으로 구성된다. 즉 앞 단 플립플롭의 변화에 따라 후단 플립플롭이 트리거되는데, 따라서 리플 카운터는 각 플립플롭의 전파지연 시간(t_d)의 총합인 T_d 시간 후에 출력이 나오게 된다. N개의 플립플롭으로 구성된 리플 카운터의 총 전파지연 시간은 다음의 식으로 표시된다.

리플 카운터의 총 전파지연 시간(T_d)	$N \times t_d$

※ t_d: 플립플롭 1단의 전파지연 시간

(1) 그림 8-8의 회로를 구성하고 출력 상태를 측정하여 표 8-4에 기록하시오.

✋ 유의사항

❶ 회로를 동작시키기 전에 리셋(Reset) 단자를 잠시 접지('0')한 후 개방시킬 것 ('1')

❷ Cp 단자에 단일 펄스를 하나씩 순차적으로 가해 가면서 각 지점의 측정된 결과를 표에 기록할 것

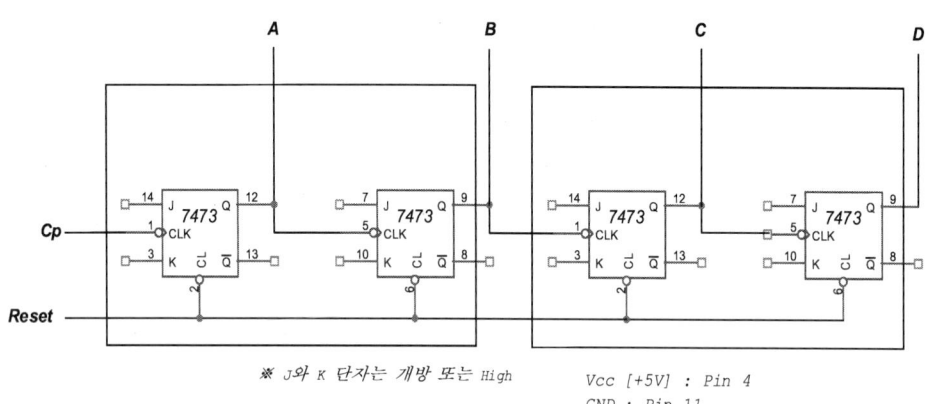

※ J와 K 단자는 개방 또는 High

Vcc [+5V] : Pin 4
GND : Pin 11

그림 8-8

표 8-4

클럭 (CLK)	출력					클럭 (CLK)	출력				
	D	C	B	A	십진수		D	C	B	A	십진수
0						16					
1						17					
2						18					
3						19					
4						20					
5						21					
6						22					
7						23					
8						24					
9						25					
10						26					
11						27					
12						28					
13						29					
14						30					
15						31					

■ 어떤 기능을 가진 회로인가?

(2) 그림 8－9의 회로를 구성하고 출력 상태를 측정하여 표 8－5에 기록하시오.

✋ 유의사항

❶ 회로를 동작시키기 전에 리셋(Reset) 단자를 잠시 접지('0')한 후 개방시킬
   것('1')

❷ Cp 단자에 단일 펄스를 하나씩 순차적으로 가해 가면서 각 지점의 측정된
   결과를 표에 기록할 것

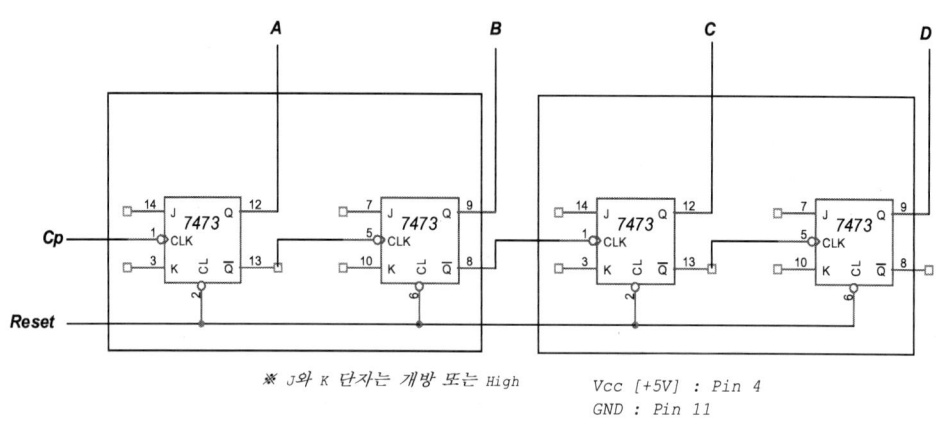

※ J와 K 단자는 개방 또는 High

Vcc [+5V] : Pin 4
GND : Pin 11

그림 8－9

표 8-5

클럭 (CLK)	출력					클럭 (CLK)	출력				
	D	C	B	A	십진수		D	C	B	A	십진수
0						16					
1						17					
2						18					
3						19					
4						20					
5						21					
6						22					
7						23					
8						24					
9						25					
10						26					
11						27					
12						28					
13						29					
14						30					
15						31					

■ 어떤 기능을 가진 회로인가?

(3) 그림 8 - 10의 회로를 구성하고 출력 상태를 측정하여 표 8 - 6에 기록하시오.

✋ 유의사항

❶ 회로를 동작시키기 전에 리셋(Reset) 단자를 잠시 접지('0')한 후 개방시킬 것('1')

❷ Cp 단자에 단일 펄스를 하나씩 순차적으로 가해 가면서 각 지점의 측정된 결과를 표에 기록할 것

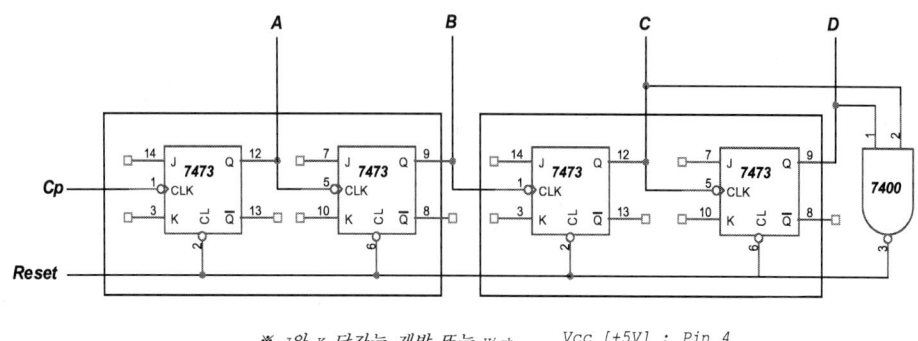

※ J와 K 단자는 개방 또는 High   Vcc [+5V] : Pin 4
                              GND : Pin 11

그림 8 - 10

표 8-6

클럭(CLK)	출력					클럭(CLK)	출력				
	D	C	B	A	십진수		D	C	B	A	십진수
0						16					
1						17					
2						18					
3						19					
4						20					
5						21					
6						22					
7						23					
8						24					
9						25					
10						26					
11						27					
12						28					
13						29					
14						30					
15						31					

■ 어떤 기능을 가진 회로인가?

실험결과 Report	학과명	학번	성명
실험 8　　비동기식 카운터			

표 8-4

클럭 (CLK)	출력					클럭 (CLK)	출력				
	D	C	B	A	십진수		D	C	B	A	십진수
0						16					
1						17					
2						18					
3						19					
4						20					
5						21					
6						22					
7						23					
8						24					
9						25					
10						26					
11						27					
12						28					
13						29					
14						30					
15						31					

■ 어떤 기능을 가진 회로인가?

표 8-5

클럭 (CLK)	출력					클럭 (CLK)	출력				
	D	C	B	A	십진수		D	C	B	A	십진수
0						16					
1						17					
2						18					
3						19					
4						20					
5						21					
6						22					
7						23					
8						24					
9						25					
10						26					
11						27					
12						28					
13						29					
14						30					
15						31					

■ 어떤 기능을 가진 회로인가?

표 8-6

클럭 (CLK)	출력					클럭 (CLK)	출력				
	D	C	B	A	십진수		D	C	B	A	십진수
0						16					
1						17					
2						18					
3						19					
4						20					
5						21					
6						22					
7						23					
8						24					
9						25					
10						26					
11						27					
12						28					
13						29					
14						30					
15						31					

■ 어떤 기능을 가진 회로인가?

연애전략은? 속공(Fast Break)일까? 지공(Set offense)일까?

| 실험 9 | 동기식 카운터 |

실험목적

- 동기식 카운터의 원리 및 사용방법을 이해하도록 한다.
- 동기식 카운터를 설계할 수 있는 능력을 배양한다.
- 카운터의 응용능력을 배양한다.

실험기기 및 재료

구분	품명	규격	수량	비고
기기	논리회로 실험장치		1	
	회로시험기		1	
	오실로스코프		1	
	주파수 카운터		1	
재료	AND 게이트	IC 74LS08	1	2입력
	AND 게이트	IC 74LS11	1	3입력
	JK FF	IC 74LS73	2	
	점퍼선		약간	

이론적 고찰

1. 동기식 카운터(Synchronous Counter)

(1) 개요

앞 장에서 실험한 비동기식 카운터에서 플립플롭의 동작은 전단(前段)의 출력에 의해 트리거되기 때문에 종속 접속하는 플립플롭의 수가 많아질수록 더 큰 전송지연이 발생하게 된다. TTL 플립플롭의 경우 하나의 전송지연 시간이 약 25ns 정도이기 때문에 플립플롭 4개를 접속한 BCD 리플 카운터의 경우 100ns정도의 전송지연 시간이 발생하게 된다. 이렇게 전송지연이 발생하게 되면 사용할 수 있는 최대 주파수를 제한하게 되며 한 상태와 다음 상태 사이에 중간 상태가 존재할 수 있어 논리상의 에러를 유발할 수도 있다.

이러한 전송지연으로 인한 문제점을 극복하고자 만들어진 카운터가 바로 동기식 카운터이다. 앞 장에서도 잠깐 언급이 있었지만 **동기식 카운터는 모든 플립플롭이 같은 클럭펄스에 의해 동시에 트리거 되도록 병렬 접속한다.** 따라서 동기식 카운터는 비동기식에 비해 전달지연 시간이 짧아진다는 장점이 있다.

(2) 동기식 카운터의 구성

동기식 카운터는 다음과 같이 구성된다.

❶ 전 플립플롭의 클럭을 공통으로 사용한다.
❷ JK 플립플롭을 사용할 때 첫 번째 플립플롭만 J, K 입력이 High 상태가 되고, 나머지 플립플롭의 J, K 입력은 다른 단(段) 플립플롭 출력의 적당한 조합에 의해 구성된다.
❸ 따라서 이 카운터는 비동기식에 비해 다음과 같은 특징이 있다.

- 동작 속도가 빠르다.
- 높은 주파수에 사용이 가능하다.
- 설계방법이 복잡하며 가격이 비싸다.

(3) 동기식 카운터의 설계방법

(가) 설계순서

❶ MOD - N 동기식 카운터를 구성하고 플립플롭의 종류 및 수를 결정한다.

❷ 여기표(勵起表: Excitation Table)를 이용하여 상태표를 작성한다.

❸ 카르노 맵을 이용하여 플립플롭의 입력 함수를 구한다.

❹ 구해진 입력 함수에 의해 회로를 구성한다.

(나) 플립플롭의 진리표와 여기표

동기식 카운터를 구성하기 위해서는 상태표를 작성해야 하는데 상태표를 작성하기 위해서는 사용되는 플립플롭에 대한 여기표를 알고 있어야 한다. 플립플롭은 주로 T형 또는 JK 플립플롭을 사용한다. 아래에서 이 두 가지 플립플롭에 대한 진리표 및 여기표를 설명한다.

표 9-1 T형 FF의 진리표

T	Q(t+T)
0	Q(t)
1	Q'(t)

표 9-2 T형 FF의 특성표

Q(t)	T	Q(t+T)
0	0	0
0	1	1
1	0	1
1	1	0

표 9-3 T형 FF의 여기표

Q(t)	Q(t+T)	T
0	0	0
0	1	1
1	0	1
1	1	0

※ 여기표: 순서논리회로를 설계할 경우 상태표를 작성하기 위해서 입력을 어떻게 해야 할 것인가를 정리하여 간단히 만든 표를 말한다.

※ 상태표: 여기표를 바탕으로 카운트 동작을 시키기 위한 입력의 요건들을 요약한 표를 말한다.

표 9-4 JK FF의 특성표

J	K	Q(t+T)
0	0	Q(t)
0	1	0
1	0	1
1	1	Q'(t)

표 9-5 JK FF의 특성표

Q(t)	J	K	Q(t+T)
0	0	0	0
0	0	1	0
0	1	0	1
0	1	1	1
1	0	0	1
1	0	1	0
1	1	0	1
1	1	1	0

표 9-6 JK FF의 여기표

Q(t)	Q(t+T)	J	K
0	0	0	x
0	1	1	x
1	0	x	1
1	1	x	0

이제 이 두 가지 플립플롭을 이용하여 동기식 카운터를 설계해 보자.

(4) MOD−8 동기식 카운터의 설계

(가) JK 플립플롭을 이용할 경우

❶ 플립플롭의 개수 결정: 3개

상태 수 $8 \leqq 2^N$에서	$N = 3$

❷ 상태표 작성: 표 9−6의 여기표를 이용

표 9−7 JK FF의 상태표

현재 상태(PS)			차기 상태(NS)			입력 함수					
Q_2	Q_1	Q_0	Q_2	Q_1	Q_0	J_2	K_2	J_1	K_1	J_0	K_0
0	0	0	0	0	1	0	x	0	x	1	x
0	0	1	0	1	0	0	x	1	x	x	1
0	1	0	0	1	1	0	x	x	0	1	x
0	1	1	1	0	0	1	x	x	1	x	1
1	0	0	1	0	1	x	0	0	x	1	x
1	0	1	1	1	0	x	0	1	x	x	1
1	1	0	1	1	1	x	0	x	0	1	x
1	1	1	0	0	0	x	1	x	1	x	1

❸ 입력 함수의 도출: 카르노 맵 이용

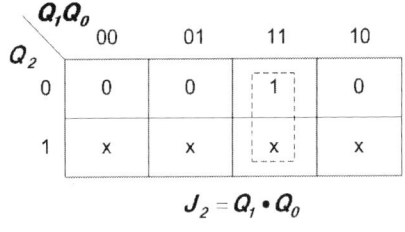

$$J_2 = Q_1 \cdot Q_0 \qquad K_2 = Q_1 \cdot Q_0$$

(a)

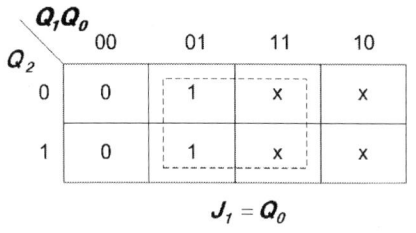

$J_1 = Q_0$

$K_1 = Q_0$

(b)

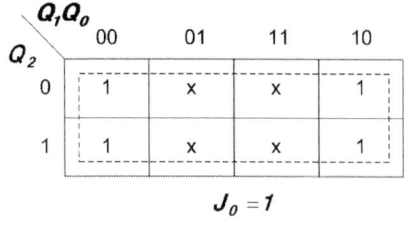

$J_0 = 1$

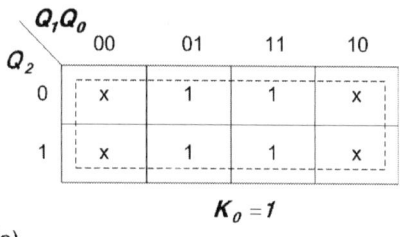

$K_0 = 1$

(c)

그림 9-1 JK 플립플롭 입력 함수 카르노 맵

❹ 회로도 작성: 입력 함수 이용

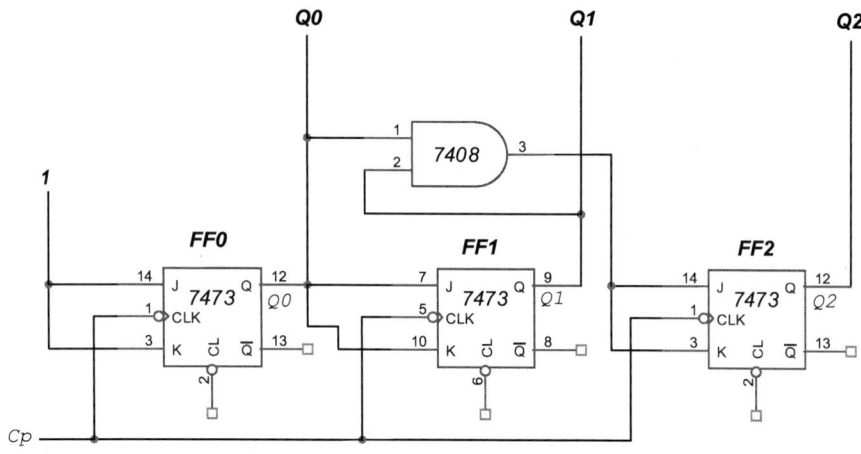

그림 9-2 MOD-8 동기식 카운터 회로

(나) T 플립플롭을 이용할 경우

❶ 플립플롭의 개수 결정: 3개

상태 수 $8 \leq 2^N$에서	$N = 3$

❷ 상태표 작성: 표 9-3의 여기표를 이용

표 9-8 T FF의 상태표

현재 상태(PS)			차기 상태(NS)			입력 함수		
Q_2	Q_1	Q_0	Q_2	Q_1	Q_0	T_2	T_1	T_0
0	0	0	0	0	1	0	0	1
0	0	1	0	1	0	0	1	1
0	1	0	0	1	1	0	0	1
0	1	1	1	0	0	1	1	1
1	0	0	1	0	1	0	0	1
1	0	1	1	1	0	0	1	1
1	1	0	1	1	1	0	0	1
1	1	1	0	0	0	1	1	1

❸ 입력 함수의 도출: 카르노 맵 이용

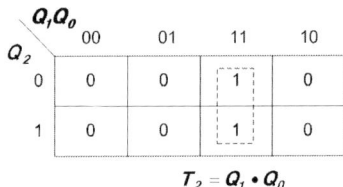

$$T_2 = Q_1 \cdot Q_0$$

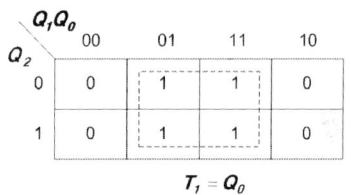

$$T_1 = Q_0$$

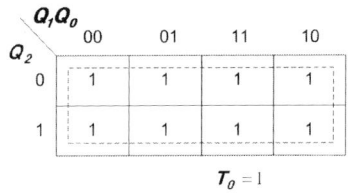

$$T_0 = 1$$

그림 9-3 T 플립플롭 입력 함수 카르노 맵

❹ 회로도 작성: 입력 함수 이용

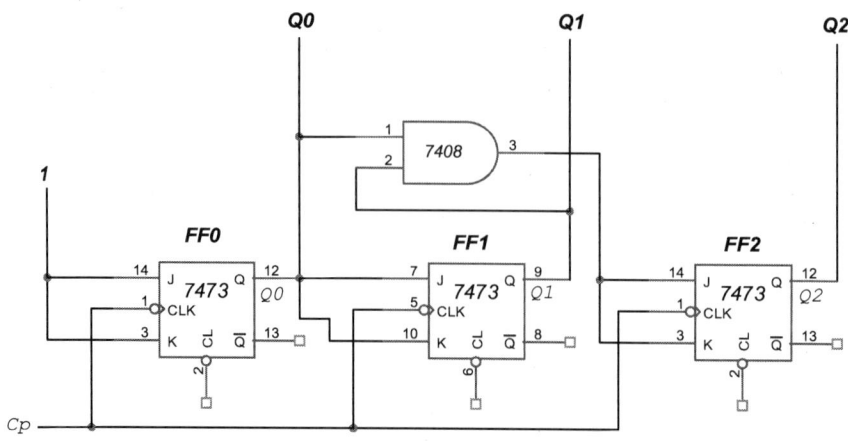

그림 9-4 MOD-8 동기식 카운터 회로

(다) MOD-8 동기식 카운터의 타이밍도

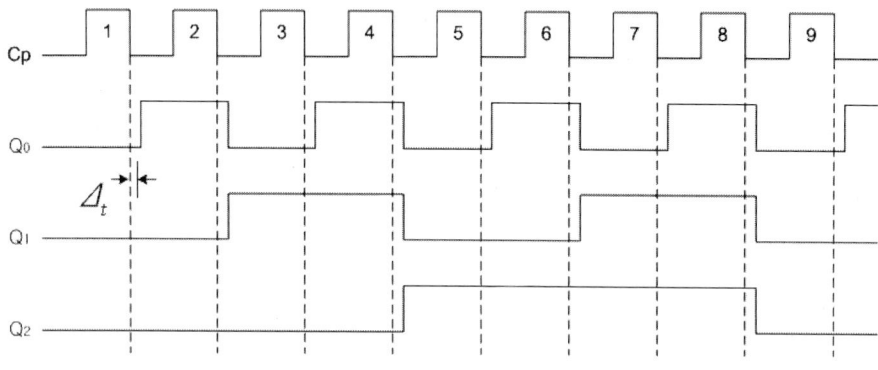

그림 9-5 MOD-8 동기식 카운터 타이밍도

(5) BCD 동기식 카운터의 설계

앞 장에서 언급한 바와 같이 BCD 카운터는 0(0000)~9(1001)까지 카운트하기 때문에 플립플롭이 4개가 필요하지만 클럭펄스 입력 10번째에서 처음의 상태로 되돌아가도록 설계하여야 한다.

❶ 플립플롭의 개수 결정: 4개

상태 수 $10 \leqq 2^N$에서	$N = 4$

❷ 상태표 작성: 표 9−6의 여기표를 이용

표 9−9 JK FF의 상태표

현재 상태(PS)				차기 상태(NS)				입력 함수							
Q_3	Q_2	Q_1	Q_0	Q_3	Q_2	Q_1	Q_0	J_3	K_3	J_2	K_2	J_1	K_1	J_0	K_0
0	0	0	0	0	0	0	1	0	x	0	x	0	x	1	x
0	0	0	1	0	0	1	0	0	x	0	x	1	x	x	1
0	0	1	0	0	0	1	1	0	x	0	x	x	0	1	x
0	0	1	1	0	1	0	0	0	x	1	x	x	1	x	1
0	1	0	0	0	1	0	1	0	x	x	0	0	x	1	x
0	1	0	1	0	1	1	0	0	x	x	0	1	x	x	1
0	1	1	0	0	1	1	1	0	x	x	0	x	0	1	x
0	1	1	1	1	0	0	0	1	x	x	1	x	1	x	1
1	0	0	0	1	0	0	1	x	0	0	x	0	x	1	x
1	0	0	1	0	0	0	0	x	1	0	x	0	x	x	1
1	0	1	0	x	x	x	x	x	x	x	x	x	x	x	x
1	0	1	1	x	x	x	x	x	x	x	x	x	x	x	x
1	1	0	0	x	x	x	x	x	x	x	x	x	x	x	x
1	1	0	1	x	x	x	x	x	x	x	x	x	x	x	x
1	1	1	0	x	x	x	x	x	x	x	x	x	x	x	x
1	1	1	1	x	x	x	x	x	x	x	x	x	x	x	x

❸ 입력 함수의 도출: 카르노 맵 이용

(a)

$Q_1 Q_0$

$Q_3 Q_2$	00	01	11	10
00	0	0	0	0
01	0	0	1	0
11	x	x	x	x
10	x	x	x	x

$$J_3 = Q_2 \cdot Q_1 \cdot Q_0$$

$Q_1 Q_0$

$Q_3 Q_2$	00	01	11	10
00	x	x	x	x
01	x	x	x	x
11	x	x	x	x
10	0	1	x	x

$$K_3 = Q_0$$

(b)

$Q_1 Q_0$

$Q_3 Q_2$	00	01	11	10
00	0	0	1	0
01	x	x	x	x
11	x	x	x	x
10	0	0	x	x

$$J_2 = Q_1 \cdot Q_0$$

$Q_1 Q_0$

$Q_3 Q_2$	00	01	11	10
00	x	x	x	x
01	0	0	1	0
11	x	x	x	x
10	0	x	x	x

$$K_2 = Q_1 \cdot Q_0$$

(c)

$Q_1 Q_0$

$Q_3 Q_2$	00	01	11	10
00	0	1	x	x
01	0	1	x	x
11	x	x	x	x
10	0	0	x	x

$$J_1 = \bar{Q}_3 \cdot Q_0$$

$Q_1 Q_0$

$Q_3 Q_2$	00	01	11	10
00	x	x	1	0
01	x	x	1	0
11	x	x	x	x
10	x	x	x	x

$$K_1 = Q_0$$

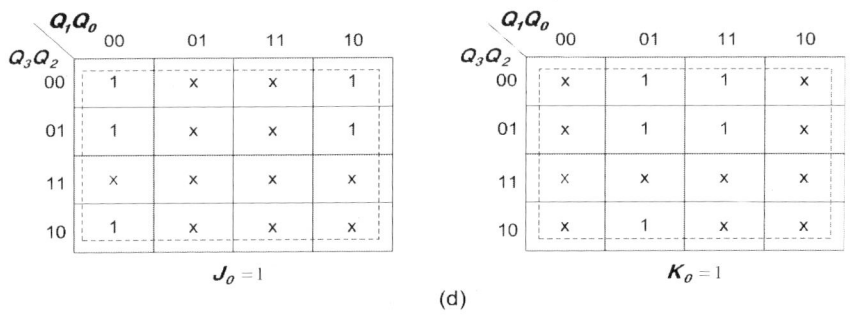

	Q_1Q_0 00	01	11	10
Q_3Q_2				
00	1	x	x	1
01	1	x	x	1
11	x	x	x	x
10	1	x	x	x

$J_0 = 1$

	Q_1Q_0 00	01	11	10
Q_3Q_2				
00	x	1	1	x
01	x	1	1	x
11	x	x	x	x
10	x	1	x	x

$K_0 = 1$

(d)

그림 9-6 JK 플립플롭 입력 함수 카르노 맵

❹ 회로도 작성: 입력 함수 이용

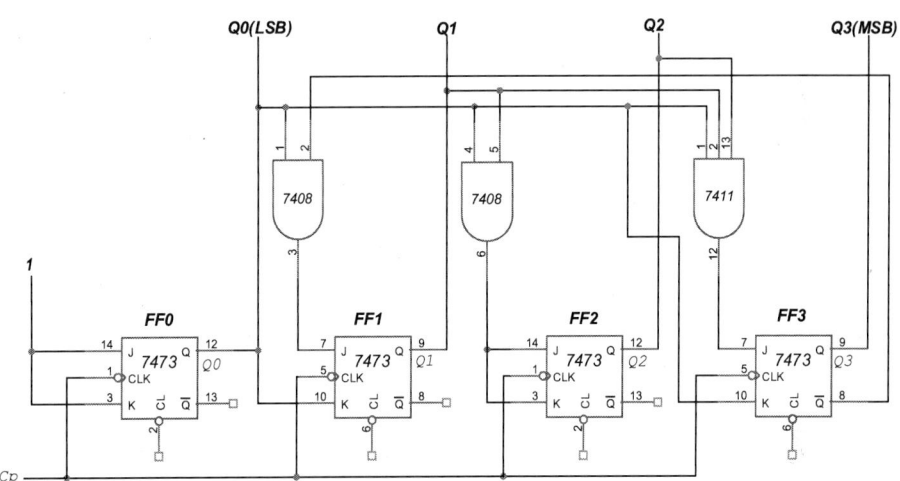

그림 9-7 BCD 동기식 카운터 회로

(1) 주어진 재료를 사용하여 MOD – 5 동기식 카운터를 설계순서에 의해 설계하고
회로를 구성하여 출력 상태를 측정하고 결과를 표 9 – 11에 기록하시오.
(재료: IC 74LS73 2개, IC 74LS08 1개)

✋ 유의사항: CLR 단자를 GND에 연결하여 ALL Clear한 후 다시 High에 고정할 것

❶ 플립플롭의 개수 결정:

❷ 상태표 작성:

표 9 – 10

현재 상태(PS)			차기 상태(NS)			입력 함수					
Q_2	Q_1	Q_0	Q_2	Q_1	Q_0	J_2	K_2	J_1	K_1	J_0	K_0

❸ 입력 함수의 도출: 카르노 맵 이용

❹ 회로도 작성: 입력 함수 이용

■ 측정결과표

표 9-11

C_p	Q_2	Q_1	Q_0	C_p	Q_2	Q_1	Q_0
0				8			
1				9			
2				10			
3				11			
4				12			
5				13			
6				14			
7				15			

(2) 그림 9-8의 회로를 구성하여 출력 상태를 측정하고 결과를 표 9-12에 기록하시오.

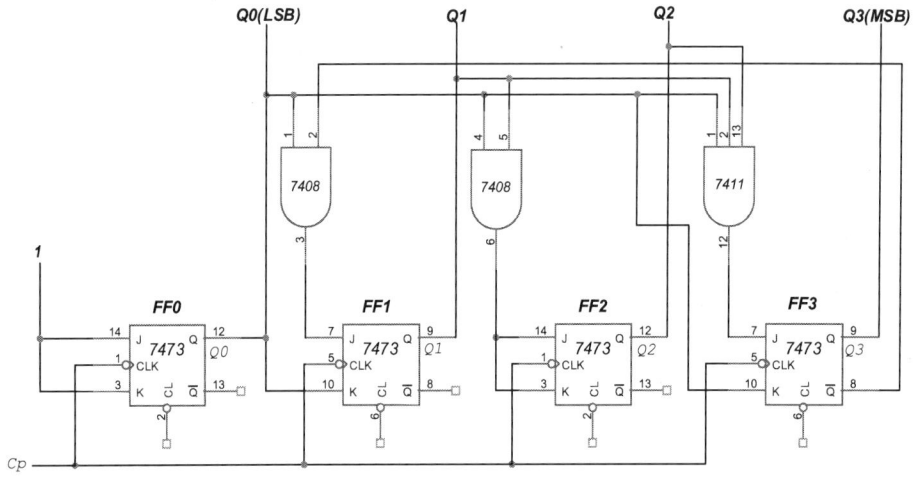

그림 9-8

■ 측정결과표

표 9-12

C_p	Q_3	Q_2	Q_1	Q_0	C_p	Q_3	Q_2	Q_1	Q_0
0					8				
1					9				
2					10				
3					11				
4					12				
5					13				
6					14				
7					15				

실험결과 Report		학과명	학번	성명
실험 9	동기식 카운터			

■ 측정결과표

표 9-11

C_p	Q_2	Q_1	Q_0	C_p	Q_2	Q_1	Q_0
0				8			
1				9			
2				10			
3				11			
4				12			
5				13			
6				14			
7				15			

■ 측정결과표

표 9 - 12

C_p	Q_3	Q_2	Q_1	Q_0	C_p	Q_3	Q_2	Q_1	Q_0
0					8				
1					9				
2					10				
3					11				
4					12				
5					13				
6					14				
7					15				

실험 10   업 · 다운 카운터

실험목적

- Up/Down 카운터의 동작원리를 이해하도록 한다.
- Up/Down 카운터를 설계할 수 있는 능력을 배양한다.
- Up/Down 카운터인 74192(74193)의 기능과 접속법을 익히도록 한다.

실험기기 및 재료

구분	품명	규격	수량	비고
기기	논리회로 실험장치		1	
	회로시험기		1	
	오실로스코프		1	
	주파수 카운터		1	
재료	AND 게이트	IC 74LS08	1	2입력
	AND 게이트	IC 74LS11	1	3입력
	OR 게이트	IC 74LS32	1	
	JK FF	IC 74LS73	2	
	동기식 Up/Down 카운터	IC 74LS193	1	Binary 카운터
	점퍼선		약간	

1. 업 · 다운 카운터(Up/Down Counter)

(1) 개요

Up/Down 카운터는 가감산(加減算) 카운터라고도 하며 2진수를 0(0000)부터 임의의 수까지 카운트하고, 반대로 임의 수에서 0(0000)까지의 수를 카운트하는 동작을 한다. 따라서 이것은 Up 카운터와 Down 카운터 동작을 공용한 카운터로서 Count Up과 Count Down 동작을 제어하는 제어입력이 필요하게 된다.

이를 이해하기 위해서 먼저 Down 카운터에 대해서 살펴보기로 하자.

(2) MOD−8 동기식 Down 카운터

Down 카운터는 지금까지 설명한 Up 카운터를 반대로 생각하면 된다. 즉 출력 상태가 0(000)부터 7(111)까지 변하는 것이 아니라 7(111)에서 0(000)으로 상태가 변하도록 구성해야 한다.

표 10−1을 참고하기 바란다.

회로 구성은 그림 10−2 MOD−8 동기식 카운터회로에서 FF 2의 입력을 Q_0와 Q_1 대신 $\overline{Q_0}$와 $\overline{Q_1}$를 AND 게이트를 통해서 연결하면 된다. 이에 관한 그림이 그림 10−2에 나타나 있다.

표 10-1 MOD-8 동기식 Down 카운터 상태표

현재 상태(PS)			차기 상태(NS)			입력 함수					
Q_2	Q_1	Q_0	Q_2	Q_1	Q_0	J_2	K_2	J_1	K_1	J_0	K_0
0	0	0	1	1	1	1	x	1	x	1	x
0	0	1	0	0	0	0	x	0	x	x	1
0	1	0	0	0	1	0	x	x	1	1	x
0	1	1	0	1	0	0	x	x	0	x	1
1	0	0	0	1	1	x	1	1	x	1	x
1	0	1	1	0	0	x	0	0	x	x	1
1	1	0	1	0	1	x	0	x	1	1	x
1	1	1	1	1	0	x	0	x	0	x	1

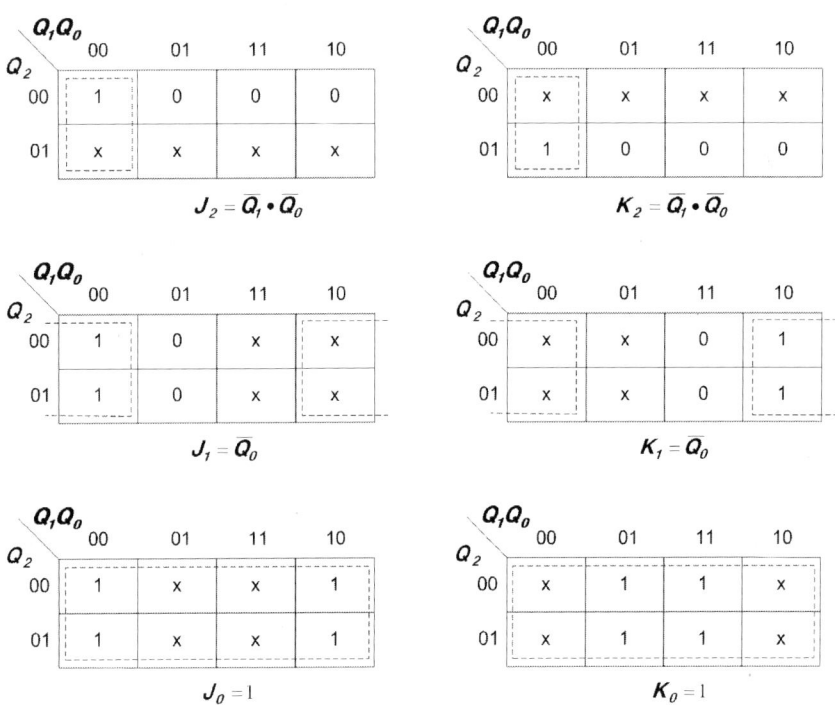

$$J_2 = \overline{Q_1} \cdot \overline{Q_0}$$

$$K_2 = \overline{Q_1} \cdot \overline{Q_0}$$

$$J_1 = \overline{Q_0}$$

$$K_1 = \overline{Q_0}$$

$$J_0 = 1$$

$$K_0 = 1$$

그림 10-1 MOD-8 Down 카운터 입력 함수 카르노 맵

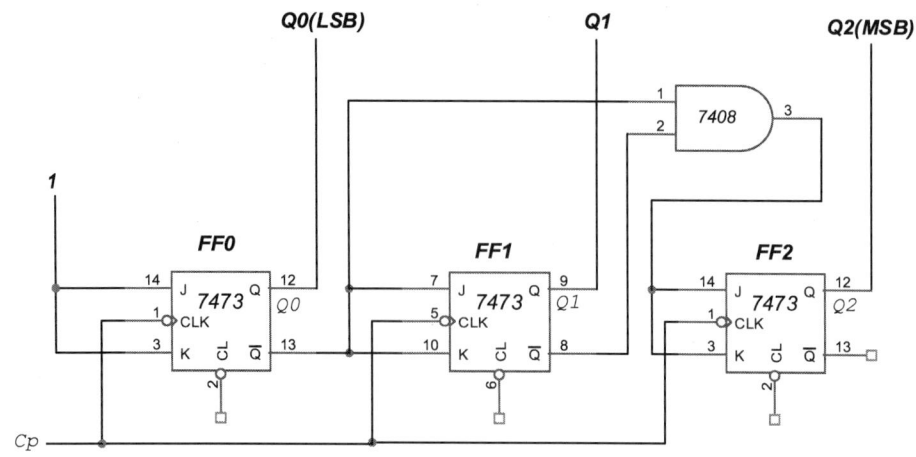

그림 10-2 MOD-8 동기식 Down 카운터 회로도

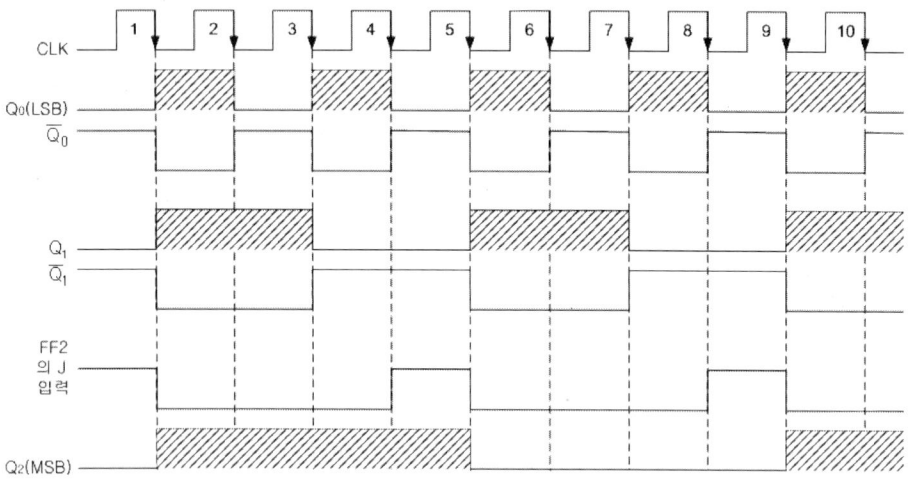

그림 10-3 MOD-8 동기식 Down 카운터 타이밍도

(3) MOD-8 동기식 Up/Down 카운터

동기식 Up/Down 카운터는 카운트 업과 카운트 다운을 명령하는 제어 입력 모두
필요하다. 이 회로는 클럭펄스의 에지(Edge) 사용에 따라 플립플롭의 정상출력(Q)
또는 반전출력($\overline{Q}$)이 다음 단 플립플롭의 J, K 입력에 공급된다. MOD-8 동기식
Up/Down 카운터의 회로도가 아래에 있다.

정상출력과 Count Up 제어입력이 High일 때 Up 카운터로 사용되고, 반전출력과 Count Down 제어입력이 High일 때 Down 카운터로 동작한다. 따라서 그림 10-4 의 카운터는

- Count Up 입력이 '1'일 때 000~111까지 카운트하고,
- Count Down 입력이 '1'일 때는 111~000까지 하향 카운트할 수 있다.
- Count Down이 '0'이고 Count Up이 '1'일 때 AND 게이트 1, 2가 동작하고 3, 4는 동작하지 않는다.
- Count Up이 '0'이고 Count Down이 '1'일 때는 반대로 AND 게이트 3, 4가 동작하고 1, 2는 동작하지 않는다.

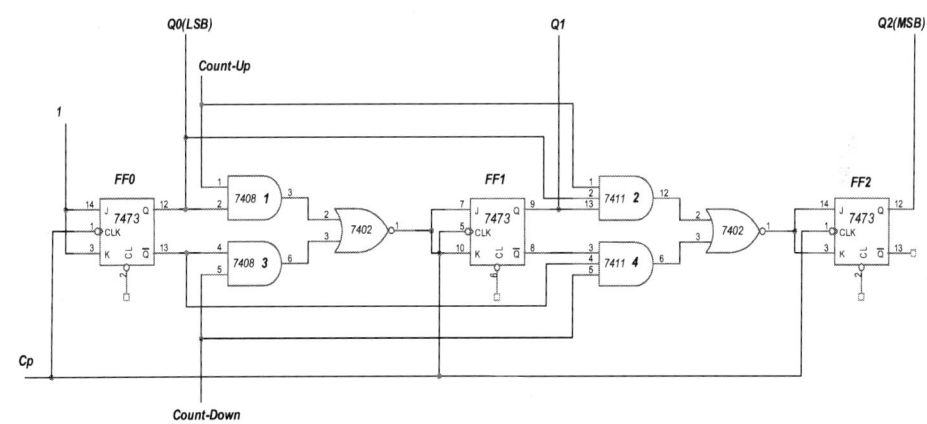

그림 10-4 MOD-8 동기식 Up/Down 카운터

대표적인 동기식 Up/Down 카운터 IC로는 74LS190과 74HC190이 있다. 아래 그림은 그 회로를 보여준다.

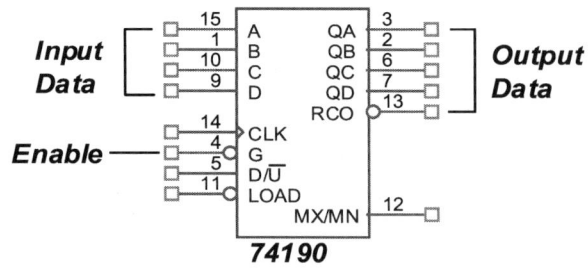

그림 10-5 IC 74LS190

(4) 프리세터블(Presettable) 동기식 Up/Down 카운터

지금까지 살펴본 카운터는 MOD가 고정된 카운터였다. 이러한 카운터는 MOD가 고정되어 있기 때문에 다양한 MOD로 활용할 수 가 없다. 따라서 '**임의의 어떤 MOD로도 변경하여 사용할 수 있는 카운터**'가 필요한데 이런 카운터가 바로 **프리세터블 카운터**이다.

74LS193은 4-비트 동기식 Up/Down Binary 카운터이다. 이 카운터는 Master Reset 입력을 가지고 있고 임의의 카운트 시작점을 미리 조절(프리세트)할 수가 있다.

표 10-2 Pin Description

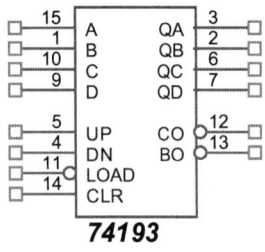

그림 10-6 IC 74LS193

Pin Name	Description
UP	Count-Up **clock** input(Rising edges)
DN	Count-Down **clock** input(Rising edges)
$\overline{LOAD}$	Asynchronous parallel load input
MR	Asynchronous master reset input
$A \sim D$	Parallel data input
$Q_a \sim Q_d$	Flip-Flop output
$\overline{CO}$	Terminal count up(Carry) output
$\overline{BO}$	Terminal count down(borrow) output

표 10 - 3 Mode Selection

MR	$\overline{LOAD}$	UP	DN	Mode
1	X	X	X	Async. Reset
0	0	X	X	Async. Preset
0	1	1	1	No change
0	1	↑	1	Count Up
0	1	1	↑	Count Down

X: Don't Care
↑: PGT(Positive going transition: 正 트리거)

각 핀의 동작 특성을 살펴보기로 하자.

❶ Up(Count - Up 클럭 입력 단자)

Up 단자에 클럭펄스 입력 시 Up 카운터로 동작

❷ DN(Count - Down 클럭 입력 단자)

DN 단자에 클럭펄스 입력 시 Down 카운터로 동작

❸ $\overline{LOAD}$(비동기 병렬입력 단자)

순간적으로 병렬입력 단자에 Low 펄스를 가하여(High에서 Low) 병렬 데이터 입력 A, B, C, D의 값을 미리 설정할 수 있다. 이 입력이 Low일 때 클럭에 관계없이 입력단의 값이 출력됨.

(단 MR 입력이 High에 있으면 영향을 미치지 못함)

❹ MR(마스터 리셋 입력 단자)

카운터 출력을 0000 상태로 만들어 주는 단자
모든 다른 입력을 무시함

❺ $Q_a \sim Q_d$(카운트 출력)

현재 카운트 상태를 출력하는 단자. Q_d는 MSB, Q_a는 LSB

❻ $\overline{CO}$(Carry 출력 단자)

카운트 출력이 최대의 값을 나타내면 이 단자가 Low가 된다. 이 단자는 $\overline{BO}$ 단자와 더불어 2개 혹은 그 이상의 74193을 연결하여 더 큰 MOD 수를 가진 다단 카운터를 구성할 때 사용한다.

❼ $\overline{BO}$(Borrow 출력 단자)

카운트 출력이 0이면 이 단자가 Low가 된다.

가. 카운트 업(Count – Up) 동작

Up 카운터로 동작하기 위해서는 Count – Up 입력에 클럭펄스가 공급되고 Down 입력은 High 상태를 유지해야 한다. 그림 10 – 7은 Up Count로 동작할 때의 타이밍도를 나타낸다.

〈동작 과정〉

① 클럭펄스 입력 1번 이전에 $\overline{LOAD}$가 Low이므로 그때의 데이터 입력 DCBA = 1000을 받아들여 클럭펄스 1번에서 Q3Q2Q1Q0 = DCBA = 1000을 출력하면서 카운트의 시작점으로 된다.

② 클럭펄스 2번의 상승 Edge에서 출력은 Q3Q2Q1Q0 = 1001이 되어 1 증가한다.

③ 차례로 1씩 증가하다가 클럭펄스 8번에서 모든 출력이 1이 되어 Q3Q2Q1Q0 = 1111이 된다.

④ 모든 출력이 1(1111)이고 클럭펄스도 Low이므로 $\overline{CO}$는 Low가 된다.

⑤ 클럭펄스 9번에서 출력은 모두 0(0000)이 되고 계속 증가하다가 MR 신호가 High가 되면 다시 모든 출력이 0(0000)이 된다.

업 카운터의 상세한 동작은 다음 타이밍도를 참고하기 바란다.

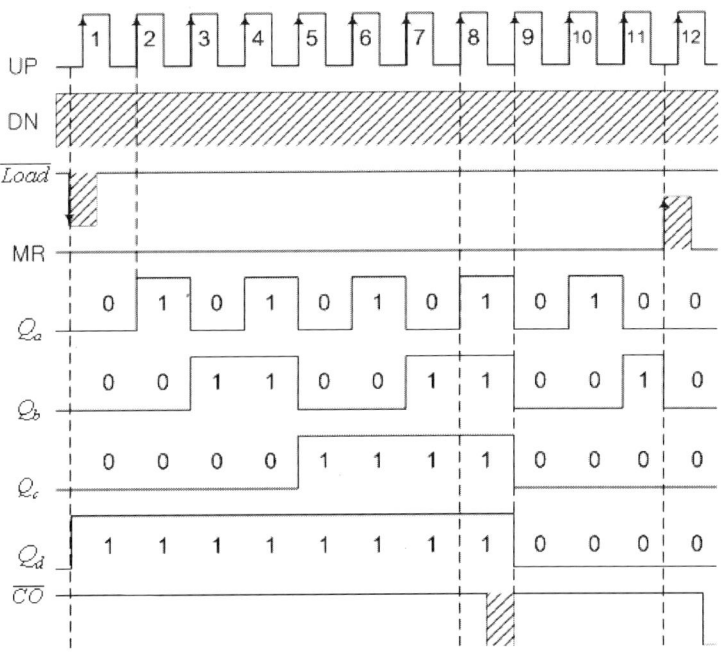

그림 10-7 Up 카운팅 시의 타이밍도

나. 카운트 다운(Count-Down) 동작

Down 카운터의 동작 조건은 Up 카운터와 반대로 Count Down 입력에 클럭펄스가 공급되고 Up 입력은 High 상태를 유지해야 한다.

〈동작 과정〉

① 클럭펄스 입력 1번 이전에는 모든 출력이 0이고 ($Q_3Q_2Q_1Q_0 = 0000$) 클럭펄스가 Low이므로 $\overline{BO}$는 Low가 된다.

② 여기서 $\overline{LOAD}$ 입력이 Low가 되어 데이터 입력 DCBA = 1000을 받아들여 클럭펄스 입력 1번에서 $Q_3Q_2Q_1Q_0 = DCBA = 1000$을 출력한다.

③ 클럭펄스 입력 2번에서 출력은 $Q_3Q_2Q_1Q_0 = 0001$이 되어 1 감소하며 클럭펄스 9번에서 모든 출력이 0(0000)이 된다.

④ 이때 모든 출력이 0(0000)이고 클럭펄스도 Low이므로 $\overline{BO}$는 Low가 된다.

⑤ $\overline{BO}$가 Low가 되면 클럭펄스 10번에서 모든 출력 $Q_3Q_2Q_1Q_0=1111$이 되어 Down Counter로서 동작하게 된다. 또한 MR 신호가 High가 되면 모든 출력값이 0(0000)이 된다.

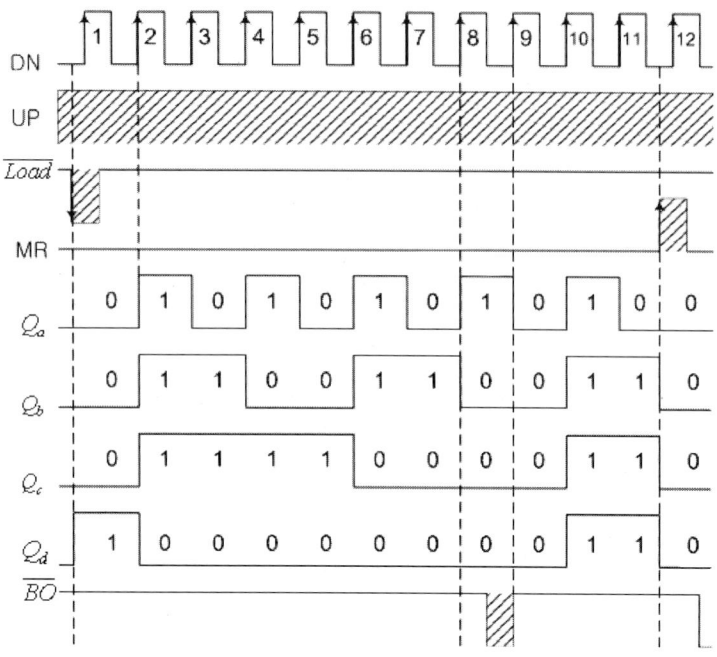

그림 10 - 8 Down 카운팅 시의 타이밍도

다. 74LS193의 응용

이제 Up/Down 카운터 74LS193이 어떻게 가변 MOD로 동작할 수 있는지 살펴보기로 하자. 이 카운터를 이용하면 별다른 논리회로를 추가하지 않고도 서로 다른 다양한 가변 MOD 수를 얻을 수 있다.

그림 10 - 9 회로는 74LS193을 이용한 MOD - 5 카운터이다. 이 회로는 0101(5)의 병렬 로드 입력을 가진 Down 카운터로 이용되고 있다. 이 회로에서 $\overline{BO}$ 출력이 다시 $\overline{LOAD}$로 feedback되고 있음에 주목하라.

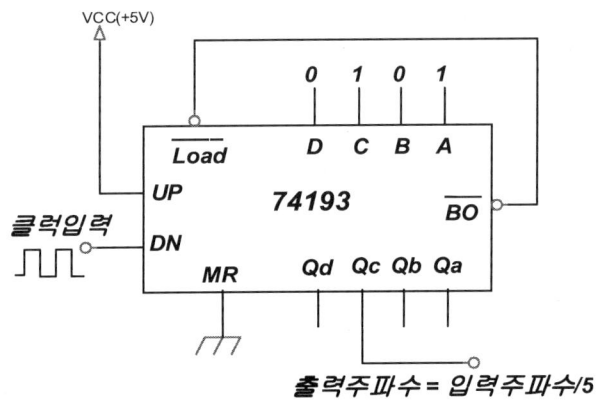

VCC(+5V)

그림 10-9 74193을 이용한 MOD-5 Down 카운터 회로도

그림 10-10은 위 회로의 타이밍도를 보여주고 있다. 이 회로가 어떻게 MOD-5 카운터로 동작하는지 살펴보기로 하자.

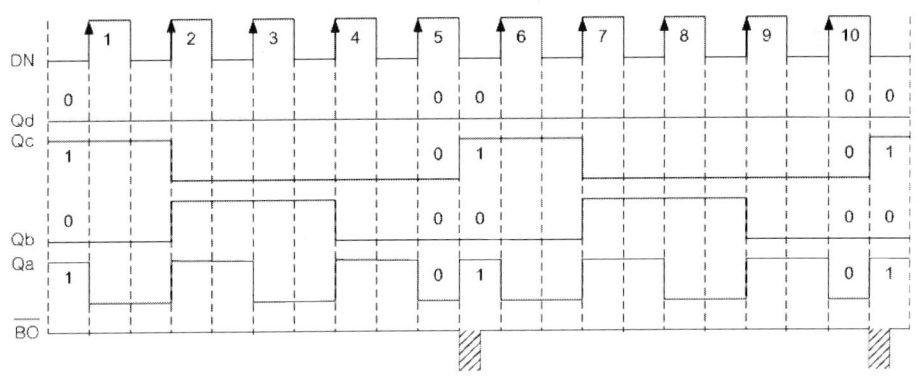

그림 10-10 74193을 이용한 MOD-5 Down 카운터의 타이밍도

위 회로의 동작 과정을 타이밍도를 이용하여 설명하기 전에 다음 상태를 가정한다. 즉 카운터가 Count Down을 하고 있고 클럭펄스 1번에서 5(0101)의 상태를 가진다고 가정한다.

〈동작 과정〉

① DN 단자로 입력되는 클럭 1번에서 클럭 5번까지는 클럭의 상승 에지에서

감소한다.

② 클릭 5번에서 출력은 0(0000)이 되고 5번 클릭이 Low로 변하는 순간 $\overline{BO}$ 단자가 Low가 된다.

③ 이때 $\overline{Load}$ 단자가 동시에 Low가 되어 카운터를 5(0101) 상태로 프리세트시킨다.

✎ $\overline{BO}$ 단자는 카운터의 출력이 $\overline{Load}$ = 0에 응답하여 5(0101) 상태로 가서 $\overline{BO}$ = 0을 유지하기 위한 조건이 제거되었기 때문에 매우 짧은 기간만 Low가 된다. 따라서 $\overline{BO}$ 에서는 글리치(Glitch) 현상이 적게 된다.

✎ 글리치(Glitch) 현상

논리 게이트의 출력에서 '**비교적 지속 시간이 짧고 원점**(原點)**이 분명치 않은 펄스 모양의 잡음이 발생하는 현상**'을 말하는 것으로 디지털 시스템에서 에러를 유발하는 원인이 되고 있다. 실험 11. 링 카운터와 존슨카운터에서 보다 자세히 살펴보기로 하겠다.

디지털 시스템의 제조 과정에서 동작 불량을 일으키는 것을 결함(defect)이라 부르며, 결함이 고장(fault)을 발생시키고, 고장 때문에 오류가 생기며, 오류는 오동작(malfunction)을 야기한다. 고장을 일으키는 결함을 가진 디지털 IC나 시스템을 찾아내고 수리하기 위한 기본적인 단계는 다음과 같이 세 가지가 있다.

❶ 결함 검출(defect detection): IC나 시스템의 동작을 올바른 동작과 비교, 관찰
❷ 결함 격리(defect isolation): IC 또는 시스템을 테스트하고 결함 부분을 격리
❸ 결함 수정(defect correction): 고장 난 부분을 교환 또는 수리

이 세 가지 단계들이 쉽게 보일지 모르지만 실제 고장 점검은 회로의 복잡성과 형태 또는 고장 기구와 가능한 고장 조사방법에 크게 의존한다.

가. 디지털 IC의 내부 결함

(1) 내부 회로의 기능 결함
내부 회로의 기능 결함이 발생되면 입력에 대한 적절한 출력 반응이 나타나지 않는다. 이 경우 대부분 내부의 어떤 요소가 고장인지를 알 수 없고, 출력값을 예측하기란 불가능하다.

(2) 입력이 내부적으로 전원이나 접지에 단락된 경우
입력이 내부적으로 전원이나 접지에 단락되는 결함은 입력 상태를 High나 Low 상태에 묶어두게 된다.

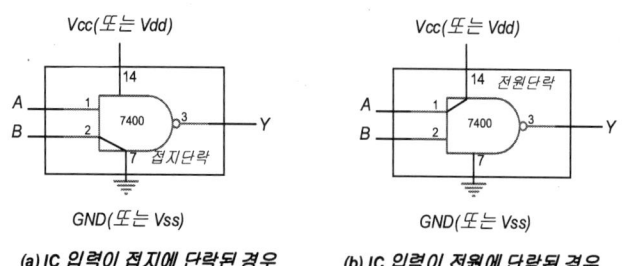

(a)IC 입력이 접지에 단락된 경우 **(b)IC 입력이 전원에 단락된 경우**

그림 10-11 입력이 전원이나 접지에 단락된 경우

　(a) 그림의 NAND 게이트 B 입력이 GND에 단락된 경우 이 게이트 출력 Y는 항상 '1' 상태가 된다. (b)의 경우 출력은 항상 B 입력의 반대 결과가 나타나게 되어 결국 NOT 게이트로 동작하게 된다.

(3) 출력이 내부적으로 전원이나 접지에 단락된 경우

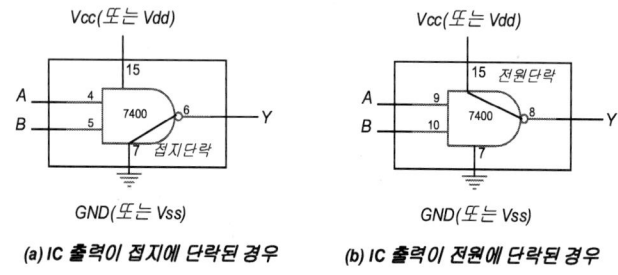

(a) IC 출력이 접지에 단락된 경우 **(b) IC 출력이 전원에 단락된 경우**

그림 10-12 출력이 전원이나 접지에 단락된 경우

　위의 경우 어떤 입력 변화가 있어도 그림 (a)의 경우 출력은 항상 '0'으로, (b)의 경우는 '1'로 나타날 것이다.

(4) 개방된 입력 또는 출력

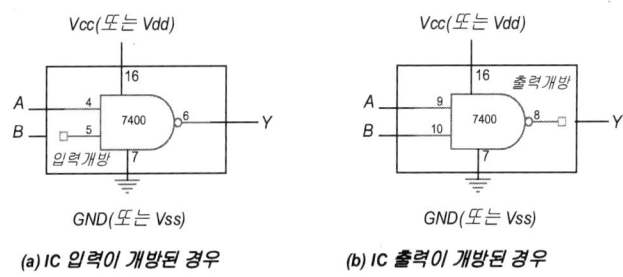

(a) IC 입력이 개방된 경우　　　**(b) IC 출력이 개방된 경우**

그림 10 – 13 입력 및 출력이 개방된 경우

(a)의 경우 TTL IC라면 개방된 입력은 항상 High 상태가 된다. CMOS IC라면 과열에 의해 IC가 손상을 입을 수도 있다. 출력이 개방된 경우는 입력을 어떻게 변화시키더라도 출력에 전압이 나타나지 않을 것이다.

(1) 주어진 재료를 사용하여 MOD-8 동기식 Up/Down 카운터를 설계하고 회로를 구성하여 출력 상태를 측정하고 결과를 표 10-6 및 표 10-7에 기록하시오. (재료: IC 74LS73 3개, IC 74LS08 2개, IC 74LS11 2개, IC 74LS32 2개)

❶ 플립플롭의 개수 결정:

❷ 상태표 작성:

표 10-4 Up 상태의 상태표

현재 상태(PS)			차기 상태(NS)			입력 함수					
Q_2	Q_1	Q_0	Q_2	Q_1	Q_0	J_2	K_2	J_1	K_1	J_0	K_0

표 10-5 Down 상태의 상태표

현재 상태(PS)			차기 상태(NS)			입력 함수					
Q_2	Q_1	Q_0	Q_2	Q_1	Q_0	J_2	K_2	J_1	K_1	J_0	K_0

❸ 입력 함수의 도출:

(가) Up 상태의 입력 함수:

$J2 =$ $K2 =$

$J1 =$ $K1 =$

$J0 =$ $K0 =$

(나) Down 상태의 입력 함수:

$J2 =$ $K2 =$

$J1 =$ $K1 =$

$J0 =$ $K0 =$

❹ 회로도 작성: 입력 함수 이용

■ 측정결과표

(가) Up Counter로 동작 시

표 10 − 6

C_p	Q_2	Q_1	Q_0	C_p	Q_2	Q_1	Q_0
0				8			
1				9			
2				10			
3				11			
4				12			
5				13			
6				14			
7				15			

(나) Down Counter로 동작 시

표 10 − 7

C_p	Q_2	Q_1	Q_0	C_p	Q_2	Q_1	Q_0
0				8			
1				9			
2				10			
3				11			
4				12			
5				13			
6				14			
7				15			

(2) 그림 10 - 14의 회로를 구성하고 출력 상태를 측정하여 결과를 표 10 - 8에 기록하시오.

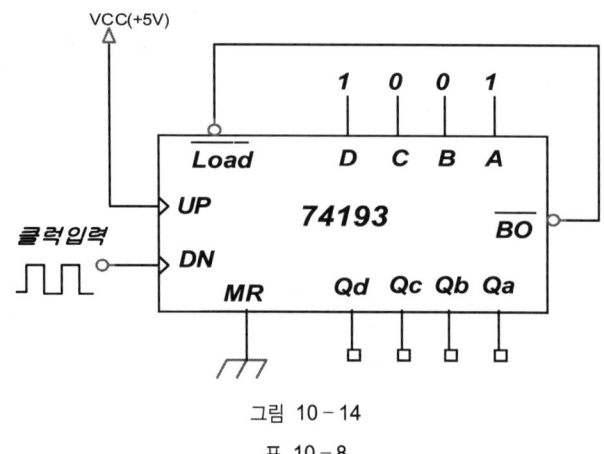

그림 10 - 14

표 10 - 8

C_p	Qd	Qc	Qb	Qq	C_p	Qd	Qc	Qb	Qa
1					11				
2					12				
3					13				
4					14				
5					15				
6					16				
7					17				
8					18				
9					19				
10					20				

실험결과 Report			학과명	학번	성명
실험 10	Up/Down 카운터				

표 10-6

C_p	Q_2	Q_1	Q_0	C_p	Q_2	Q_1	Q_0
0				8			
1				9			
2				10			
3				11			
4				12			
5				13			
6				14			
7				15			

표 10-7

C_p	Q_2	Q_1	Q_0	C_p	Q_2	Q_1	Q_0
0				8			
1				9			
2				10			
3				11			
4				12			
5				13			
6				14			
7				15			

표 10 - 8

C_p	Qd	Qc	Qb	Qa	C_p	Qd	Qc	Qb	Qa
1					11				
2					12				
3					13				
4					14				
5					15				
6					16				
7					17				
8					18				
9					19				
10					20				

실험 11 링 카운터와 존슨 카운터

실험목적

- 링 카운터(Ring Counter)의 기본 원리를 이해하도록 한다.
- 자체 시동(Self starting)이 가능한 링 카운터의 구성방법을 익힌다.
- 존슨 카운터(Johnson counter)의 원리를 이해하고 구성방법을 익힌다.

실험기기 및 재료

구분	품명	규격	수량	비고
기기	논리회로 실험장치		1	
	회로시험기		1	
	오실로스코프		1	
	주파수 카운터		1	
재료	NAND 게이트	IC 74LS00	1	2입력
	NAND 게이트	IC 74LS10	1	3입력
	JK FF	IC 74LS76	2	
	점퍼선		약간	

1. 링 카운터(Ring Counter)

(1) 개요

앞 장에서 설명한 시프트레지스터를 이용하여 카운터 회로를 구성할 수도 있다. 시프트레지스터를 이용한 카운터를 시프트레지스터 카운터라고 하며 이러한 카운터에는 링 카운터(Ring Counter)와 존슨 카운터(Johnson Counter)가 있다. 이들 카운터는 모두 시프트레지스터를 구성하는 마지막 단(段) 플립플롭의 입력으로 궤환(feedback)시켜 구성한다.

(2) 시동펄스(Initiate pulse)가 필요한 링 카운터

링 카운터는 '**마지막 단 플립플롭의 정상출력(Q)을 처음 단 플립플롭의 입력으로 feedback시킨 시프트레지스터 회로**'를 말하며 **순환 시프트레지스터**라고도 한다. 링 카운터의 기본 구조는 그림 11 – 2와 같다.

일반적으로 2진 카운터나 시프트레지스터는 N개의 플립플롭으로 구성되어 있으면 2^N개의 상태를 가질 수 있으나 링 카운터는 N개의 상태밖에는 가질 수 없다. 즉 MOD – 8 2진 카운터를 구성 시 **MOD 수**=2^N(N: 플립플롭의 수)에서 플립플롭이 3개 반 있으면 되나 MOD – 8 링 카운터는 **MOD 수**=N(N: 플립플롭의 수)에서 플립플롭이 8개가 필요하게 된다. 이와 같이 링 카운터는 MOD 수만큼 플립플롭이 필요하므로 비효율적이라 말할 수 있지만 다음과 같은 장점 때문에 많이 사용되고 있는 카운터이다.

❑ 링 카운터의 장점
❶ 디코딩 신호를 플립플롭의 출력에서 직접 얻기 때문에 카운터 신호를 디코딩하는 다른 논리회로를 첨가하지 않아도 된다.

(표 11 - 1 링 카운터 상태천이표 참조)

❷ 카운터 신호를 이용하여 시스템을 제어할 경우 디코딩 게이트를 사용하지 않음으로 글리치(Glitch) 현상이 발생치 않는다.

❸ 즉 조합논리회로가 필요 없이 순서논리회로만으로도 모든 기능을 수행할 수 있다.

표 11 - 1 상태천이표

입력	출력				
클럭	Q_4	Q_3	Q_2	Q_1	Q_0
0	1	0	0	0	0
1	0	1	0	0	0
2	0	0	1	0	0
3	0	0	0	1	0
4	0	0	0	0	1
5	1	0	0	0	0
6	0	1	0	0	0
7	0	0	1	0	0
...	...	...	...	...	...

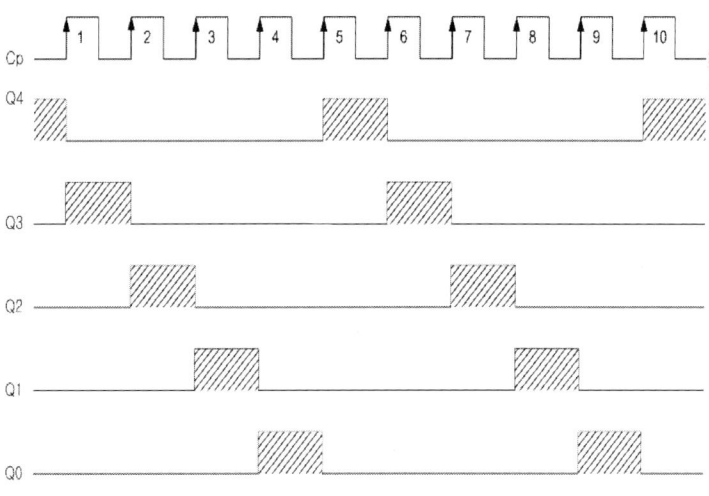

그림 11 - 1 MOD - 5 링 카운터 출력 타이밍도

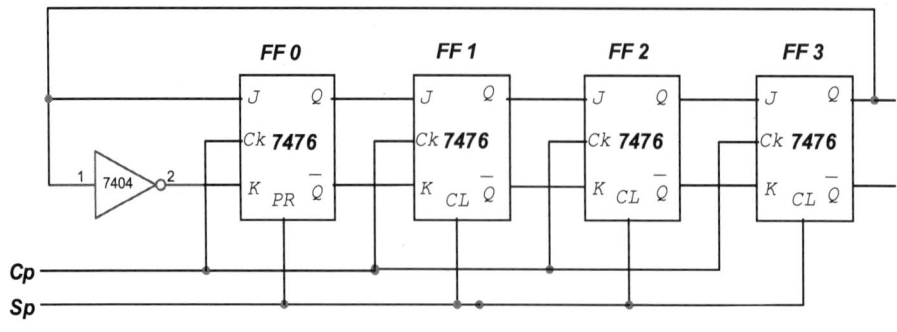

그림 11 - 2 시동펄스가 필요한 링 카운터

위 그림에서 **Sp 단자**는 **시동펄스** 단자인데 이 단자의 접속 중 FF 0에만 프리세트(PR) 단자로 접속되고 나머지는 클리어(CLR) 단자에 접속된 것에 유의할 것

❏ 동작 과정

❶ 시동펄스를 잠시 '0'으로 한 후 다시 '1'로 하면 FF 0만 '1'이 되고 나머지는 '0' 상태가 된다.

❷ 이때 Cp 단자에 펄스를 가하면 '1'의 상태가 순차적으로 다음 플립플롭으로 이동한다.

❸ 4개의 클럭이 가해지면 FF 0만 '1'이 되는 본래의 상태로 돌아온다.

❹ 이때 '0'을 순차적으로 회전할 때는 시동펄스 단자의 접속을 FF 0만 CLR 단자로 하고 나머지는 PR 단자로 하면 된다.

이러한 종류의 링 카운터는 반드시 시동펄스가 있어야만 동작하며 전원이 나가거나 외부에서 잡음이 있을 경우에도 사용되지 않는 상태가 나타나지 않게 하려면 반드시 시동펄스를 가해야만 하는 불편한 점이 있다.

🔲 알고 넘어갑시다

〈글리치 현상〉

글리치 현상이란 '비교적 지속 시간이 **짧고 원점(原點)이 분명치 않은 펄스 모양의 잡음(雜音)**'을 말하며 주로 D/A 변환기에서 입력 디지털 양이 변할 때 출력에 과도적인 스파이크(Spike)나 오버 슈트(Over shoot)가 생기는 일을 말한다.

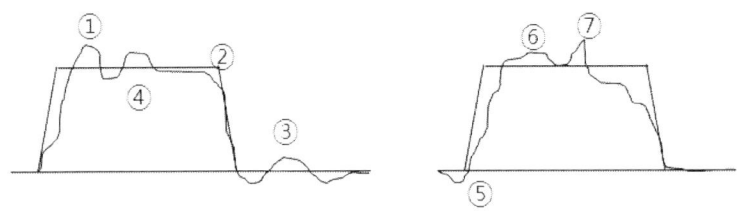

그림 11 - 3 잡음의 종류별 형태

❶ 오버 슈트: 펄스의 평균 폭을 넘는 잡음

❷ 각결(角缺): 펄스의 모서리를 깎는 잡음

❸ 기저선 통과: 펄스의 기준선에서 발생하는 잡음

❹ 링깅(Ringing): 적은 잡음이 계속 발생하는 현상

❺ 프레슈트: 펄스의 전단에서 반대 방향으로 발생하는 잡음

❻ 후크(Hook): 전화기의 Hook와 같이 올라갔다 내려오는 형상의 잡음

❼ 스파이크: 펄스의 일시적, 순간적인 과도현상

(3) 시동펄스(Initiate pulse)가 필요 없는 링 카운터

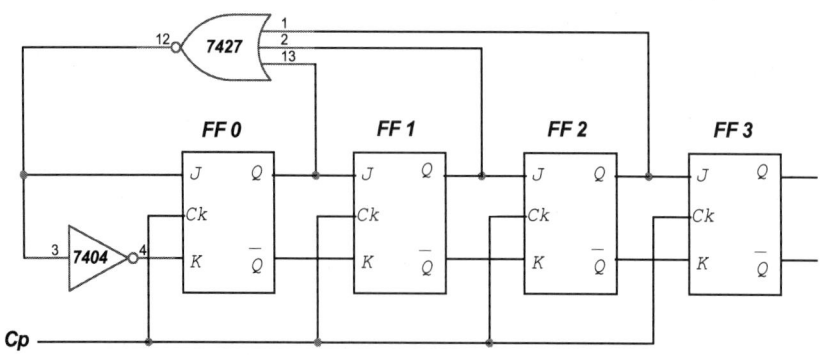

그림 11-4 시동펄스가 필요 없는 링 카운터

위 그림은 앞에서 언급된 카운터처럼 시동펄스가 필요 없는 링 카운터 회로이다. 이 카운터는 자체 시동(Self starting)이 가능한데 최종 단을 제외한 전 플립플롭의 출력을 NOR 게이트를 통하여 입력으로 궤환시킴으로써 가능하게 된다.

그림 11-5는 앞의 회로에서 가질 수 있는 모든 가능한 상태를 보여주고 있다.

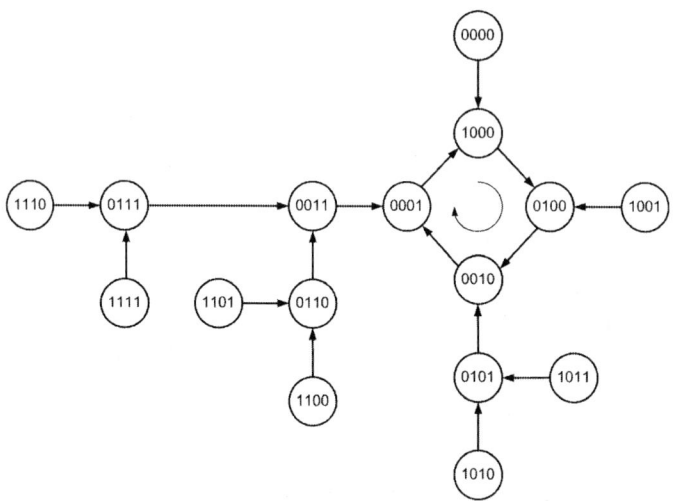

그림 11-5 자체 시동이 가능한 4비트 링 카운터의 상태도

그림 11-4에서 각 플립플롭의 출력 단자가 NOR 게이트의 입력으로 연결되어 있기 때문에 출력값이 3개 중 하나라도 '1'이면 NOR 게이트의 출력은 '0'이 된다. 물론 NOR 게이트에 연결된 플립플롭의 출력이 모두 $0(Q_3Q_2Q_1Q_0 = 0000)$일 경우는 곧바로 1000의 상태로 바뀌게 된다. 그림 11-5는 1000→0100→0010→0001→1000으로 '1'이 우측으로 계속 순환하는 카운터이다. 이 그림에서 초기 상태가 다음과 같이 각각 주어진다면 클럭펄스가 입력될 때마다 상태가 어떻게 변하는지 살펴보자.

❶ 초기 상태가 '$Q_3Q_2Q_1Q_0 = 1110$'인 경우

1110 → 0111 → 0011 → 0001 → 1000 → 0100 → 0010 → 0001 →

❷ 초기 상태가 '$Q_3Q_2Q_1Q_0 = 1111$'인 경우

1111 → 0111 → 0011 → 0001 → 1000 → 0100 → 0010 → 0001 →

❸ 초기 상태가 '$Q_3Q_2Q_1Q_0 = 1101$'인 경우

1101 → 0110 → 0011 → 0001 → 1000 → 0100 → 0010 → 0001 →

❹ 초기 상태가 '$Q_3Q_2Q_1Q_0 = 1100$'인 경우

1100 → 0110 → 0011 → 0001 → 1000 → 0100 → 0010 → 0001 →

❺ 초기 상태가 '$Q_3Q_2Q_1Q_0 = 1010$'인 경우

1010 → 0101 → 0010 → 0001 → 1000 → 0100 → 0010 → 0001 →

❻ 초기 상태가 '$Q_3Q_2Q_1Q_0 = 1011$'인 경우

1011 → 0101 → 0010 → 0001 → 1000 → 0100 → 0010 → 0001 →

❼ 초기 상태가 '$Q_3Q_2Q_1Q_0 = 1001$'인 경우

1001 → 0100 → 0010 → 0001 → 1000 → 0100 → 0010 → 0001 →

위의 경우를 보면 초기 상태가 어떤 상태에 있었더라도 클럭펄스가 4번만 가해지면 링 카운터에 필요한 상태가 되도록 되어 있다. 따라서 시동펄스로 프리세트 또는 리세트시킬 필요가 없게 된다. 그러나 카운터에 필요한 초기 값을 스스로 갖기 위해서는 클럭펄스 주기의 4배에 해당되는 시간이 필요하게 된다. 즉 이 회로는 수정 시간이 문제시되지 않는 경우에 사용이 가능하며 '0'을 회전시키는 링 카운터를 만들 경우에는 NOR 게이트 대신 NAND 게이트를 사용하면 된다.

2. 존슨 카운터(Johnson Counter)

존슨 카운터는 링 카운터와 다른 점이 피드백시키는 값이 다르다는 것이다. 링 카운터는 마지막 단 플립플롭의 정상출력(Q)을 피드백시키지만 존슨 카운터는 반전출력 ($\overline{Q}$)을 피드백시킨다. 존슨 카운터는 모든 플립플롭을 리세트 상태로 만들므로 이 회로에 클럭이 가해지면 피드백 값이 '1'이 된다.

MOD 수는 플립플롭의 수가 N개라면 MOD 수=2×N이 되어 항상 플립플롭의 2배가 된다. 존슨 카운터는 링 카운터보다 플립플롭을 적게 사용하지만 링 카운터에서는 필요 없었던 디코딩(Decoding) 회로를 필요로 한다. 그러나 일반 카운터에 비해서는 대단히 간단한 형태로 구성할 수 있다.

예를 들어 디코딩 회로가 내장되어 있는 10진 카운터를 설계하는 경우를 생각해 보자.

❶ 일반 MOD-N 카운터로 설계할 경우
 - 플립플롭은 4개, 디코딩 회로에 4입력 AND 게이트 10개가 필요하다.

❷ 존슨 카운터로 설계할 경우
 - 플립플롭은 5개가 있어야 하나 디코딩 회로에는 2입력 AND 게이트 10개만 있으면 가능하다.

❸ 일반 링 카운터로 설계할 경우
 - 디코딩 회로는 필요 없으나 플립플롭이 10개가 필요하다.

따라서 이런 경우 존슨 카운터로 설계하는 것이 가장 바람직하다.

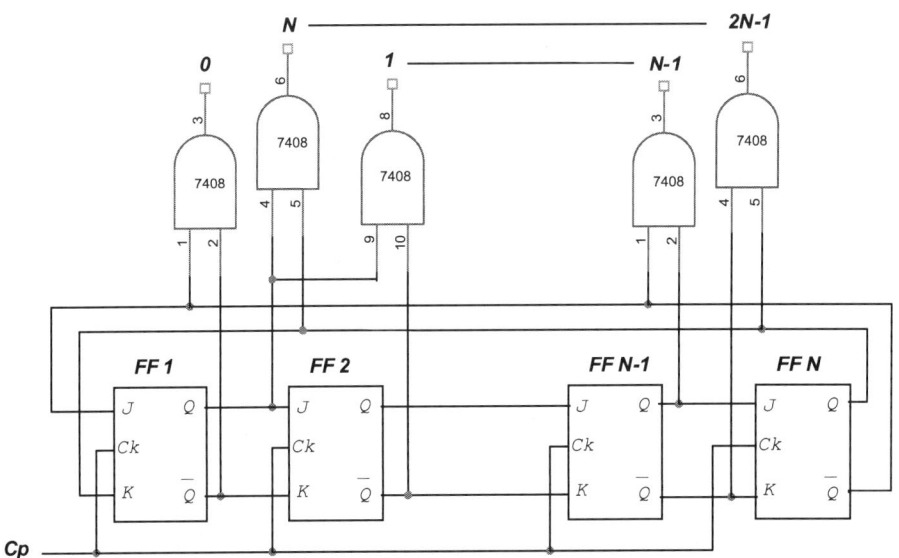

그림 11 - 6 디코더를 첨가한 존슨 카운터

· 표 11 - 2 존슨 카운터의 상태천이표

Cp	플립플롭 번호						디코더 논리식
	1	2	3	··············	N−1	N	
0	0	0	0	··············	0	0	$Q'_1 \cdot Q'_N$
1	1	0	0	··············	0	0	$Q_1 \cdot Q'_2$
2	1	1	0	··············	0	0	$Q_2 \cdot Q'_3$
3	1	1	1	··············	0	0	$Q_3 \cdot Q'_4$
····	····	····	····	··············	····	····	····
N−1	1	1	1	··············	1	0	$Q_{N-1} \cdot Q'_N$
N	1	1	1	··············	1	1	$Q_1 \cdot Q_N$
N+1	0	1	1	··············	1	1	$Q'_1 \cdot Q_2$
N+2	0	0	1	··············	1	1	$Q'_2 \cdot Q_3$
N+3	0	0	0	··············	1	1	$Q'_3 \cdot Q_4$
····	····	····	····	··············	····	····	····
2N−1	0	0	0	··············	0	1	$Q'_{N-1} \cdot Q_N$
2N	0	0	0	··············	0	0	$Q'_1 \cdot Q'_N$

위의 그림과 상태표를 보면 존슨 카운터의 동작 상태를 이해할 수가 있다.

클럭펄스가 가해질 때마다 변하는 출력 상태를 고려해 보자.

클럭이

♣ 없을 때는
- 0번 AND 게이트가 '1'이 되도록 하기 위해 $Q'_1 \cdot Q'_N$의 출력을

♣ 1번 입력되면
- 1번 AND 게이트가 '1'이 되도록 하기 위해 $Q_1 \cdot Q'_2$의 출력을

♣ 2번 입력되면
- 2번 NAD 게이트가 '1'이 되도록 하기 위해 $Q_2 \cdot Q'_3$의 출력을

......

♣ N-1번 입력되면
- N-1번 AND 게이트가 '1'이 되도록 하기 위해 $Q_{N-1} \cdot Q'_N$의 출력을

♣ N번 입력되면
- N번 AND 게이트가 '1'이 되도록 하기 위해 $Q_1 \cdot Q_N$의 출력을

♣ 2N-1번 입력되면
- 2N-1번 AND 게이트가 '1'이 되도록 하기 위해 $Q'_{N-1} \cdot Q_N$의 출력을

♣ 2N번 입력되면
- 2N번 게이트가 '1'이 되도록 하기 위해 $Q'_1 \cdot Q'_N$의 출력을

각각의 AND 게이트에 연결하여 AND 게이트가 차례로 상태 '1'을 나타내도록 한다. 따라서 AND 게이트의 출력은 디코더(Decoder)의 역할을 수행하게 되는 것이다.

선수교체! 나는 들어가는 사람일까? 나오는 사람일까?

(1) 다음의 지시에 따라 실험을 진행하시오.

❶ 그림 11-7의 회로를 구성하고 +5V 전압을 가(加)하시오.

❷ 표 11-3에 있는 조건 Ⓐ에 주어진 대로 플립플롭 세트 또는 리세트시켜 놓을 것
♣ 세트(Set): 각 IC의 PR 단자를 잠시 접지('0')시킨 후 개방('1')시킨다.
　 리세트(Reset): 각 IC의 CLR 단자를 잠시 접지('0')시킨 후 개방('1')시킨다.

❸ 단일 펄스 발생기로 클럭펄스를 1개식 순차적으로 인가하면서 표 11-3에 지시된 각 지점의 전압을 측정하여 표에 기록하시오.

❹ 표 11-3의 조건 Ⓑ, Ⓒ에 대해서도 위의 과정을 반복하시오.

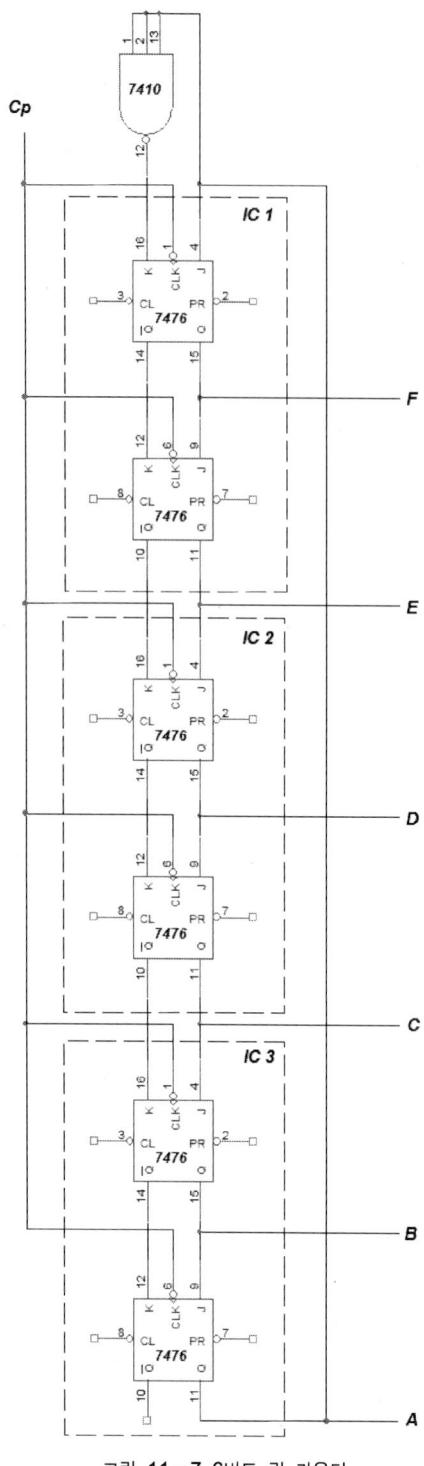

그림 11-7 6비트 링 카운터

표 11-3

조건	Cp	IC 1		IC 2		IC 3	
		F	E	D	C	B	A
Ⓐ	0	0	+5V	0	+5V	0	0
	1						
	2						
	3						
	4						
	5						
	6						
	7						
Ⓑ	0	+5V	0	0	+5V	+5V	0
	1						
	2						
	3						
	4						
	5						
	6						
	7						
Ⓒ	0	+5V	0	0	0	0	0
	1						
	2						
	3						
	4						
	5						
	6						
	7						

(2) 다음의 지시에 따라 실험을 진행하시오.

❶ 그림 11-8의 회로를 구성하고 +5V 전압을 가(加)하시오.

❷ 표 11-4에 있는 조건 Ⓐ에 주어진 대로 플립플롭 세트 또는 리세트시켜 놓을 것

❸ 단일 펄스 발생기로 클럭펄스를 1개씩 순차적으로 인가하면서 표 11-4에 지시된 각 지점의 전압을 측정하여 표에 기록하시오.

❹ 표 11-4의 조건 Ⓑ에 대해서도 위의 과정을 반복하시오.

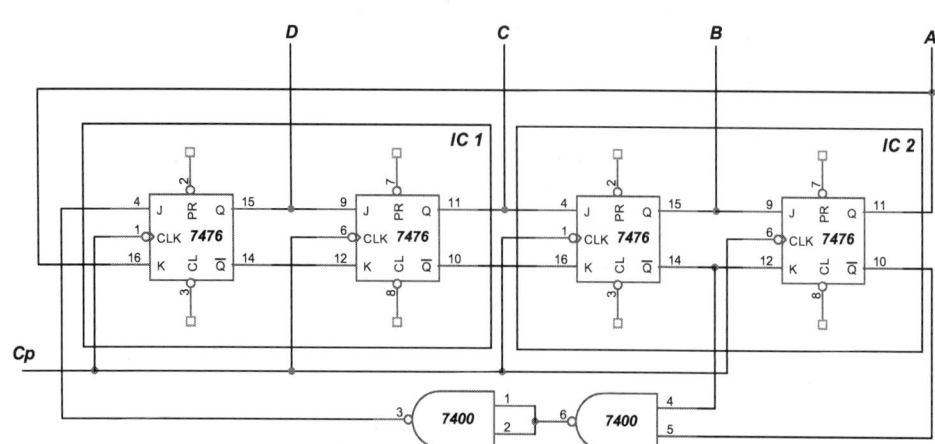

그림 11-8 자체 시동이 가능한 존슨 카운터

표 11-4

조건	Cp	IC 1		IC 2		조건	Cp	IC 1		IC 2	
		D	C	B	A			D	C	B	A
Ⓐ	0	0	0	+5V	0	Ⓑ	0	0	0	0	0
	1						1				
	2						2				
	3						3				
	4						4				
	5						5				
	6						6				
	7						7				
	8						8				
	9						9				

표 11-3

조건	Cp	IC 1		IC 2		IC 3	
		F	E	D	C	B	A
Ⓐ	0	0	+5V	0	+5V	0	0
	1						
	2						
	3						
	4						
	5						
	6						
	7						
Ⓑ	0	+5V	0	0	+5V	+5V	0
	1						
	2						
	3						
	4						
	5						
	6						
	7						
Ⓒ	0	+5V	0	0	0	0	0
	1						
	2						
	3						
	4						
	5						
	6						
	7						

표 11-4

조건	Cp	IC 1		IC 2		조건	Cp	IC 1		IC 2	
		D	C	B	A			D	C	B	A
Ⓐ	0	0	0	+5V	0	Ⓑ	0	0	0	0	0
	1						1				
	2						2				
	3						3				
	4						4				
	5						5				
	6						6				
	7						7				
	8						8				
	9						9				

실험 12   시프트레지스터

실험목적

● 시프트레지스터(Shift register)의 기본 원리를 이해하도록 한다.
● 각종 시프트레지스터의 구성방법과 용도를 익히도록 한다.
● 플립플롭을 이용한 회로의 응용력을 향상시킨다.

실험기기 및 재료

구분	품명	규격	수량	비고
기기	논리회로 실험장치		1	
	회로시험기		1	
	오실로스코프		1	
	주파수 카운터		1	
재료	NAND 게이트	IC 74LS00	2	2입력
	NOT 게이트	IC 74LS04	1	
	JK FF	IC 74LS76	2	
	점퍼선		약간	

1. 시프트레지스터(Shift Register)

(1) 레지스터 개요

플립플롭은 1비트의 데이터를 저장할 수 있는 기억요소로 활용될 수 있다고 배웠다. 이 플립플롭을 여러 개 사용하면 여러 비트의 데이터를 저장할 수 있다. n비트 2진수 데이터를 저장하기 위해서는 플립플롭이 n개가 필요하다. 즉 4비트 2진수 데이터를 저장하는 데는 플립플롭이 4개가 필요하다는 말이다. 이것을 **레지스터**라고 한다.

다시 말해 레지스터는 '**데이터를 일시적으로 저장하여 디지털 회로와 입출력 장치를 연결하고, 보수연산, 곱셈, 나눗셈과 같은 산술연산을 할 수 있는 디지털 회로**'라고 할 수 있다. 레지스터로는 T형 플립플롭을 제외한 모든 플립플롭이 사용될 수 있으나 주로 RS와 D 플립플롭이 많이 사용되고 있다.

'0'과 '1'이 2진 데이터를 레지스터로 전달하는 방법은 직렬이동(Serial shifting)과 병렬이동(Parallel shifting) 두 가지가 있는데 전자(前者)는 한 번에 1비트씩 직렬로 이동시키는 방법을, 후자(後者)는 모든 비트를 동시에 레지스터로 이동시키는 방법을 말한다.

(2) 시프트레지스터 개요

n비트의 시프트레지스터는 n개의 플립플롭을 정렬시켜 데이터를 상호간에 일정한 방향으로 1비트씩 이동시키는 레지스터이다. 레지스터에서의 시프트 동작은 클럭펄스에 의하여 레지스터 내의 한 단에서 다른 단의 플립플롭으로 또는 레지스터의 내부나 외부로의 데이터 이동을 하는 것을 말한다. 여기서는 시프트 동작의 원

리와 시프트레지스터의 종류에 대해 살펴보기로 하겠다.

(가) 시프트 동작의 원리

그림 12-1은 D 플립플롭 3개를 이용한 3비트 시프트레지스터로 각 플립플롭은 클럭펄스의 상승 에지(Edge)에서 동작한다.

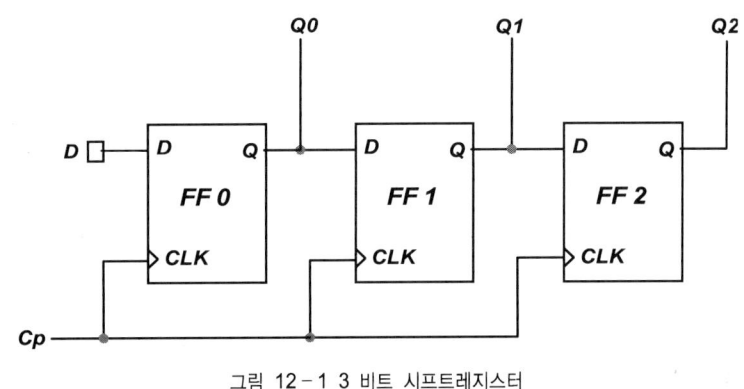

그림 12-1 3 비트 시프트레지스터

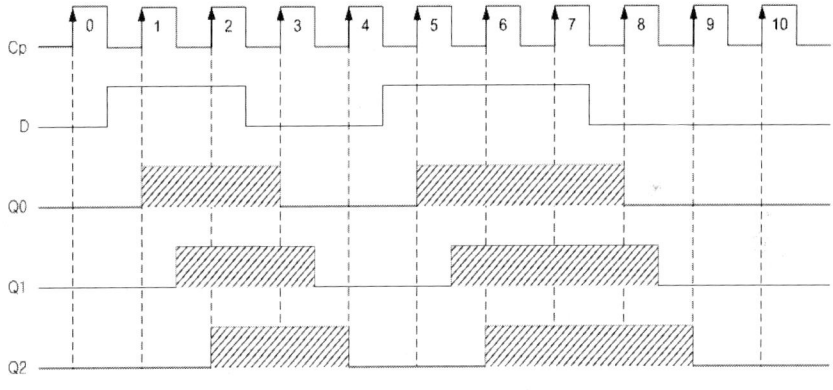

그림 12-2 3비트 시프트레지스터 타이밍도

플립플롭 0(FF 0)의 D 입력이 위의 타이밍도와 같이 주어진다고 할 때 시프트레지스터의 동작 과정을 살펴보자.

〈동작 과정〉

❶ 클럭펄스 1번의 상승 에지에서 입력 D는 FF 0의 출력 Q0＝D가 되고

❷ 클럭펄스 2번의 상승 에지에서 Q0는 FF 1로 시프트하여 출력 Q1에 Q0의 값이 나타난다.

❸ Q1은 클럭 3번의 상승 에지에서 FF 2로 시프트하여 출력 Q2에 Q0의 값을 전달한다.

따라서 D 입력은 클럭펄스의 상승 에지 시간 간격으로 1비트씩 오른쪽으로 차례로 시프트(shift)됨을 알 수 있다.

(나) 시프트레지스터의 종류

시프트레지스터는 데이터의 입력과 출력의 형태에 따라 다음과 같이 4가지로 구분할 수 있다.

❶ 직렬입력 – 직렬출력(SISO) 레지스터
❷ 직렬입력 – 병렬출력(SIPO) 레지스터
❸ 병렬입력 – 직렬출력(PISO) 레지스터
❹ 병렬입력 – 병렬출력(PIPO) 레지스터

위 레지스터들의 동작 형태를 그림으로 나타내면 다음과 같다.

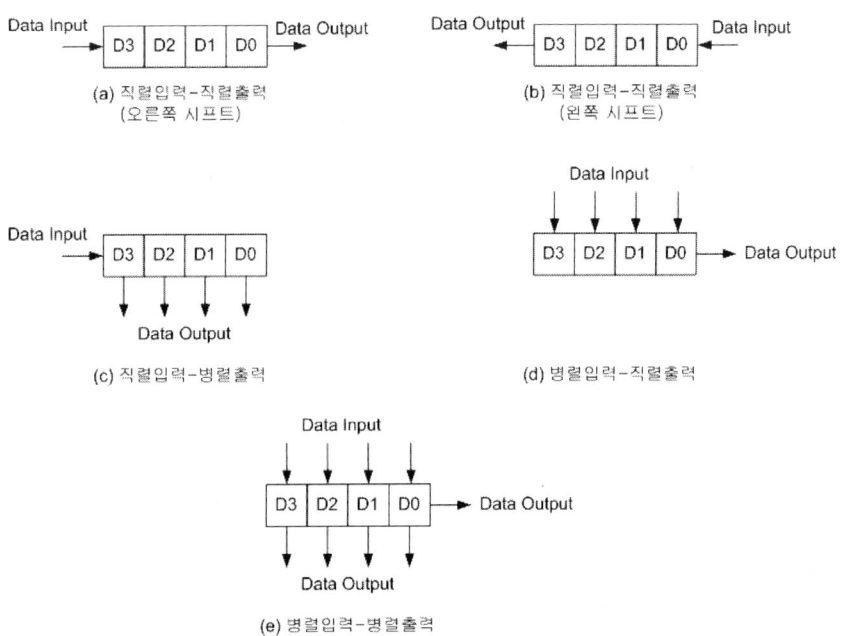

(a) 직렬입력-직렬출력
(오른쪽 시프트)

(b) 직렬입력-직렬출력
(왼쪽 시프트)

(c) 직렬입력-병렬출력

(d) 병렬입력-직렬출력

(e) 병렬입력-병렬출력

그림 12 - 3 시프트레지스터의 입, 출력 동작

(3) 시프트레지스터의 설계

(가) 우측 시프트레지스터

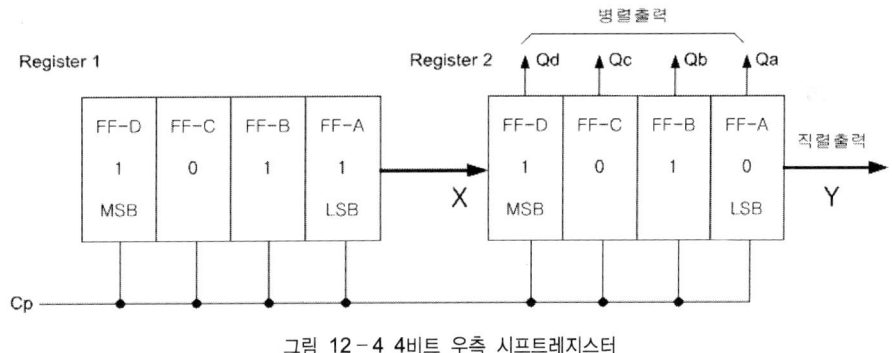

그림 12 - 4 4비트 우측 시프트레지스터

표 12-1 4비트 우측 시프트레지스터의 상태표

Cp	시각	레지스터 1 (4비트 직렬입력)				레지스터 2 (기억값)				직렬 출력 (Y)
		D	C	B	A	D	C	B	A	
0	t_n	1	0	1	1	1	0	1	0	X
1	t_{n+1}	X	1	0	1	1	1	0	1	0
2	t_{n+2}	X	X	1	0	1	1	1	0	1
3	t_{n+3}	X	X	X	1	0	1	1	1	0
4	t_{n+4}	X	X	X	X	1	0	1	1	1

█████ : 4비트 병렬출력 데이터

표 12-2 Qd 값의 진리표

tn		tn+1	입력 함수	
X	Qd	Qd	Jd	Kd
0	0	0	0	X
0	1	0	X	1
1	0	1	1	X
1	1	1	X	0

표 12-3 Qc 값의 진리표

tn		tn+1	입력 함수	
Qd	Qc	Qc	Jc	Kc
0	0	0	0	X
0	1	0	X	1
1	0	1	1	X
1	1	1	X	0

표 12-4 Qb 값의 진리표

tn		tn+1	입력 함수	
X	Qd	Qd	Jd	Kd
0	0	0	0	X
0	1	0	X	1
1	0	1	1	X
1	1	1	X	0

표 12-5 Qa 값의 진리표

tn		tn+1	입력 함수	
Qd	Qc	Qc	Jc	Kc
0	0	0	0	X
0	1	0	X	1
1	0	1	1	X
1	1	1	X	0

그림 12-4의 Qd, Qc, Qb, Qa에 대한 진리표는 표 12-2~5와 같이 나타낼 수 있다. 따라서 각단의 플립플롭 입력값은 다음과 같이 구할 수가 있다.

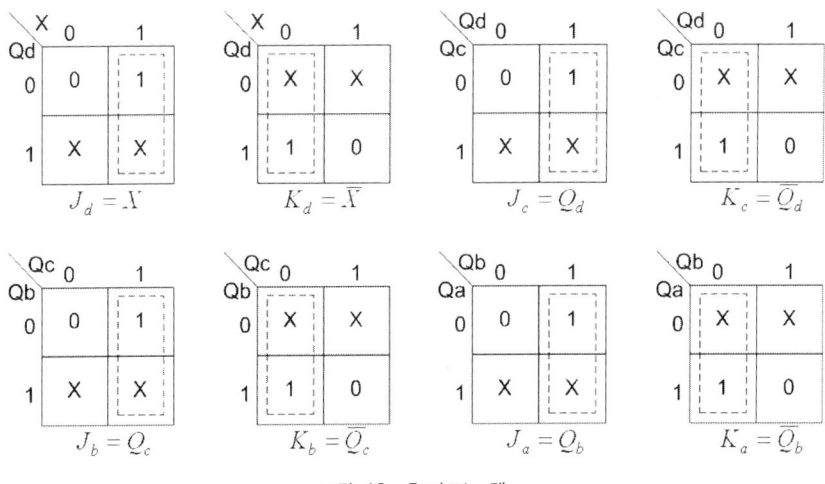

그림 12-5 카르노 맵

입력 함수값에 의해 4비트 우측 시프트레지스터 회로도를 구성하면 아래와 같다.

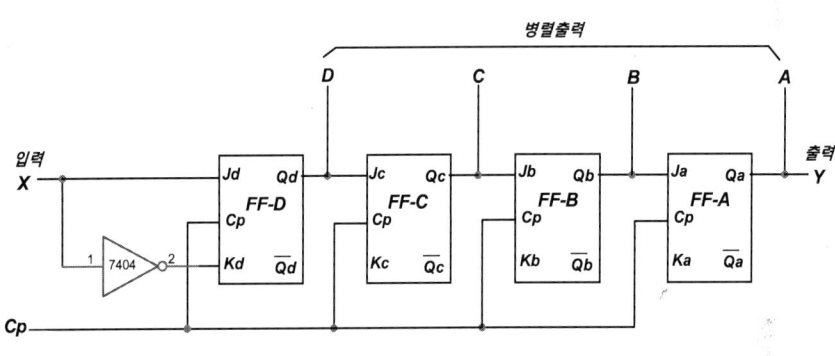

그림 12-6 JK 플립플롭 동기식 4비트 우측 시프트레지스터

(나) 좌측 시프트레지스터

좌측 시프트레지스터는 레지스터의 최하위(LSB)에 입력을 넣어 플립플롭에 기억된 정보를 윗자리로 이동시키도록 하여야 한다. 따라서 우측 시프트레지스터와 반대로 생각하면 된다. 설계는 같은 요령으로 하면 되고 회로는 다음과 같다.

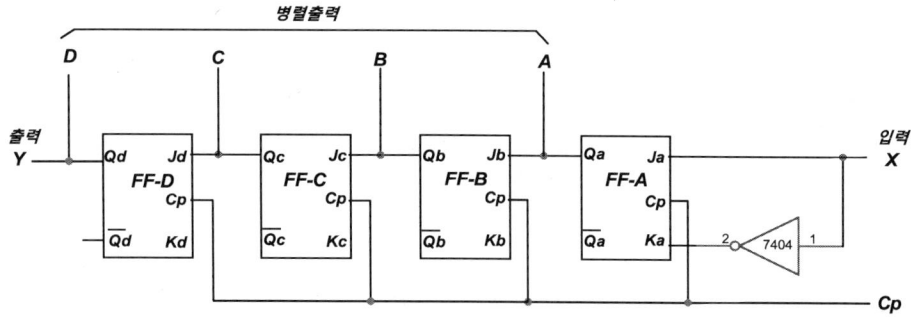

그림 12-7 JK 플립플롭 동기식 4비트 우측 시프트레지스터

(다) 병렬 입출력형(Parallel input/output) 시프트레지스터

n비트의 직렬 시프트레지스터는 데이터를 전송하는 데 n비트의 시간만큼 소요가
되므로 고속 동작을 요하는 시스템에서는 적합지 못하다. 따라서 1 비트의 시간으
로 n비트의 데이터를 이동시킬 수 있는 시프트레지스터가 필요한데 이것이 바로
병렬 입출력형 시프트레지스터이다. 그림 12-8에 이 회로를 나타내었다. 이 회로
는 프리세트(PR) 단자와 클리어(CLR) 단자가 모두 있는 JK 플립플롭을 이용하여
구성할 수 있다.

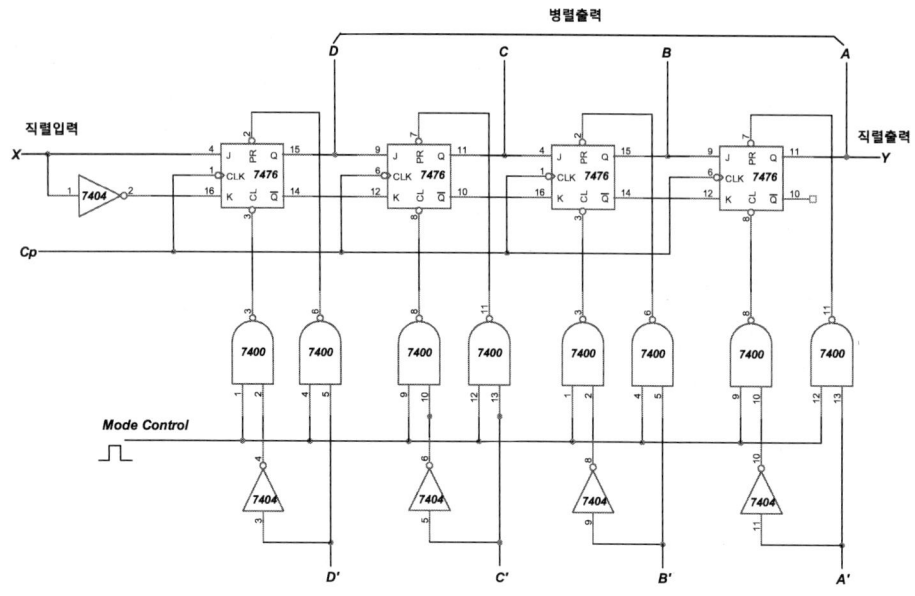

그림 12-8 병렬 입출력형 시프트레지스터

〈동작 과정〉

❶ 4비트의 병렬 입출력 단자(D'C'B'A')에 데이터(0 또는 1)를 가하고

❷ mode control 단자에 '1'을 넣어 주면
- 데이터 입력이 '1'인 경우는 우측 NAND 게이트의 출력이 '0'이 되어
  F－F을 프리세트(PR)시켜 Q를 '1'로 하고,
- 데이터 입력이 '0'인 경우는 좌측 NAND 게이트의 출력이 '0'이 되어
  F－F을 리셋(CLR)시켜 Q를 '0'으로 하여
- 각 입력 단자에 가해진 데이터가 각 F－F의 DCBA 단자로 출력된다.

❸ mode control 단자에 '0'을 넣어 주면
- 각 F－F의 PR 및 CLR 단자가 '1'이 되므로 병렬데이터 입력을 F－F에 넣을
  수가 없게 된다.

❹ Cp에 가해지는 클럭펄스는 위 그림이 우측 시프트레지스터이므로 F－F에 저
  장된 데이터를 우측으로 시프트시켜 Y 단자에서 직렬출력을 얻을 수도 있다.

따라서 이러한 레지스터는 직렬입력을 직렬 또는 병렬출력으로, 병렬입력을 직렬
또는 병렬출력으로 보낼 수 있게 된다.

(1) 그림 12 – 9의 회로를 구성하고 지시에 따라 출력 상태를 측정하여 결과를 표
12 – 6 및 12 – 7에 기록하시오.

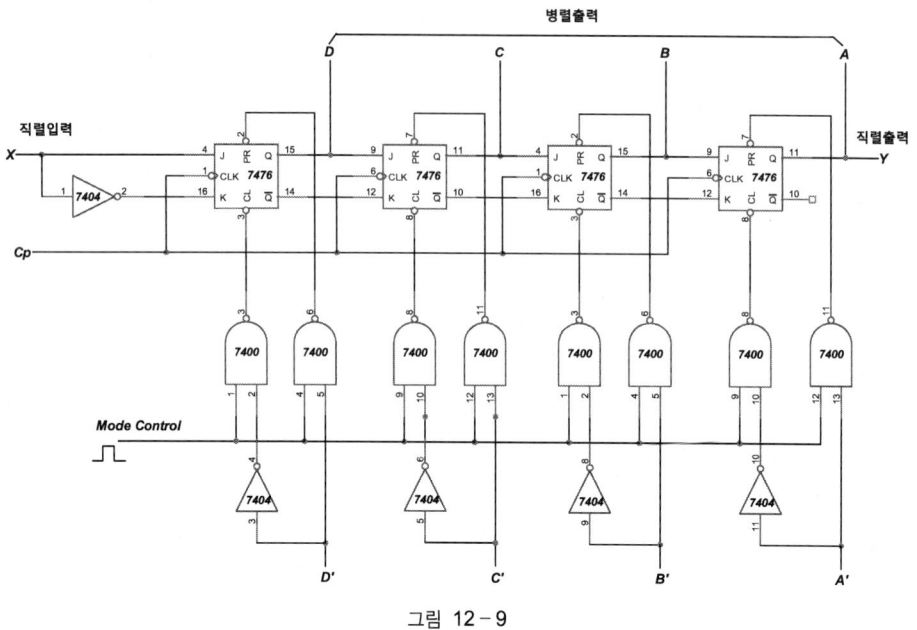

그림 12 – 9

반드시 아래에 지시하는 순서대로 실험할 것

〈병렬 입출력 실험〉

❶ mode control 단자를 ‘1’로 한다.

❷ 표 12 – 6의 순서에 의해 병렬입력 데이터를 설정한 후 클럭펄스를 가(加)할 것

표 12-6

Cp	입력				출력			
	D'	C'	B'	A'	D	C	B	A
1	0	0	0	0				
2	0	0	0	1				
3	0	0	1	0				
4	0	0	1	1				
5	0	1	0	0				
6	0	1	0	1				
7	0	1	1	0				
8	0	1	1	1				
9	1	0	0	0				
10	1	0	0	1				
11	1	0	1	0				
12	1	0	1	1				
13	1	1	0	0				
14	1	1	0	1				
15	1	1	1	0				
16	1	1	1	1				

〈직렬 입출력 실험〉

❶ mode control 단자를 '1'로 하고 입력 데이터를 모두 '0'으로 해서 모든 F-F
을 클리어시킨다.

❷ mode control 단자를 '0'으로 고정한다.

❸ 표 12-7의 순서에 의해 직렬입력 데이터를 설정 후 클럭펄스를 가(加)할 것

표 12-7

Cp	입력	출력	F-F 출력				Cp	입력	출력	F-F 출력			
	X	Y	D	C	B	A		X	Y	D	C	B	A
1	0						17	0					
2	0						18	1					
3	0						19	0					
4	0						20	0					
5	0						21	0					
6	0						22	1					
7	0						23	0					
8	1						24	1					
9	0						25	0					
10	0						26	1					
11	1						27	1					
12	0						28	0					
13	0						29	0					
14	0						30	1					
15	1						31	1					
16	1						32	1					

실험결과 Report		학과명	학번	성명
실험 12	시프트레지스터			

표 12 - 6

Cp	입력				출력			
	D'	C'	B'	A'	D	C	B	A
1	0	0	0	0				
2	0	0	0	1				
3	0	0	1	0				
4	0	0	1	1				
5	0	1	0	0				
6	0	1	0	1				
7	0	1	1	0				
8	0	1	1	1				
9	1	0	0	0				
10	1	0	0	1				
11	1	0	1	0				
12	1	0	1	1				
13	1	1	0	0				
14	1	1	0	1				
15	1	1	1	0				
16	1	1	1	1				

표 12-7

Cp	입력	출력	F-F 출력				Cp	입력	출력	F-F 출력			
	X	Y	D	C	B	A		X	Y	D	C	B	A
1	0						17	0					
2	0						18	1					
3	0						19	0					
4	0						20	0					
5	0						21	0					
6	0						22	1					
7	0						23	0					
8	1						24	1					
9	0						25	0					
10	0						26	1					
11	1						27	1					
12	0						28	0					
13	0						29	0					
14	0						30	1					
15	1						31	1					
16	1						32	1					

뒤로 처졌다고 생각되는가? 지금이 바로 백어택(Back Attack)을 시도할 찬스다!

제3부   응용회로 제작

실험 13   D/A 변환

실험목적

- D/A 변환기의 기본 원리와 의미 및 필요성을 이해하도록 한다.
- D/A 변환을 위해 필요한 741형 OP AMP의 원리를 이해하도록 한다.
- D/A 변환기의 구성방법을 익히도록 한다.

실험기기 및 재료

구분	품명	규격	수량	비고
기기	논리회로 실험장치		1	
	회로시험기		1	
	오실로스코프		1	
	직류전원공급장치		1	±15V, +5V
재료	Counter	IC 74LS93	1	4Bits Banary Cnt
	741형 OP AMP	μ 741	1	
	저항	10kΩ	2	
	저항	20kΩ	1	
	저항	30kΩ	1	
	저항	40kΩ	1	
	저항	80kΩ	1	
	가변저항 VR	10kΩ	1	
	점퍼선		약간	

1. D/A 변환(Digital to analog converter)

(1) 개요

디지털 신호를 제어장치에 이용하는 경우, 즉 예를 들면 지진계의 서버 모터 (Servo-motor)를 제어하는 경우 또는 용광로의 압력이나 온도 등을 제어하기 위해 가열기(Heater)를 구동시키는 경우 등에는 정보를 처리하기 위해서 디지털 시스템을 사용하게 되는데 실제로 디지털 시스템의 출력은 2진값으로 되어 있기 때문에 이에 대응되는 아날로그값으로 바꾸어 주어야 할 경우가 있다. 이렇게 디지털값을 아날로그값으로 바꾸어 주는 과정을 D/A 변환기라고 하고 그러한 역할을 하는 변환기를 D/A 변환기(Digital to analog converter)라고 한다. 즉 D/A 변환기는 디지털 기기에서 결과를 얻기 위한 디코딩 장치(Decoding device)라고 할 수 있다.

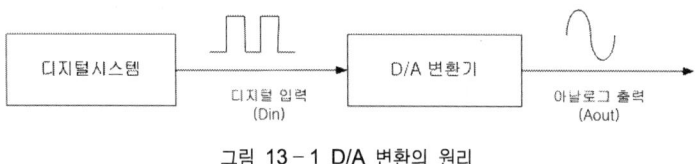

그림 13-1 D/A 변환의 원리

위 그림에서 D/A 변환기의 출력값을 Aout라 하고, 디지털 입력신호값을 Din이라 할 때 아날로그 출력값은 다음 공식에 의해 구할 수 있다.

$$A_{out} = K \times D_{in}$$

K는 비례상수로 디지털 입력의 변화에 의해 발생되는 아날로그 출력의 가장 적은 변화량(즉 전압 등을 분해할 수 있는 세밀도의 정도)을 나타내는 D/A 변환기의 분해능(Resolution)을 말한다. 즉 분해능이 8비트인 경우 Full Scale의 $1/2^8(1/256)$로

분해할 수 있는데 이 값을 말한다.

좀 더 자세히 살펴보기로 하자.

2진 디지털 입력신호를 등가의 아날로그 출력신호로 바꾸는 데는 각각의 디지털 레벨을 하나의 2진식으로 가중된(Weighed) 전압(또는 전류)으로 바꾸어야 한다. 2진식 등가 가중치(Binary equivalent weight)란 현재의 디지털값이 디지털 자리에 해당하는 가중치의 전체 합에 비해 얼마를 차지하느냐 하는 것이다.

예를 들어 설명해 보겠다.

표 13 - 1

2^2	2^1	2^0	전압 환산값
0	0	0	0 V
0	0	1	1 V
0	1	0	2 V
0	1	1	3 V
1	0	0	4 V
1	0	1	5 V
1	1	0	6 V
1	1	1	7 V

표 13 - 2

Bit	가중치
2^0	1/7
2^1	2/7
2^2	4/7
합	7/7

표 13 - 1의 3비트 2진 진리표를 보자. 이 진리표에서 8개의 가능한 신호를 등가의 아날로그 전압으로 바꾼다고 생각해 보자. 가장 작은 디지털값 '000'을 0V라고 하고, 가장 큰 값 '111'을 7V라고 한다면, 이 범위가 바로 만들어지는 아날로그 신호의 범위가 된다. 이 진리표에서 가장 작은 자리, 즉 2^0은 아날로그 출력값에서 전체 눈금의 1/7과 같은 변화를 일으키게 한다. 따라서 저항형(전압) 분할기로 D/A를 설계할 경우 2^0 위치에 있는 '1'이 1V가 되게 설계해야 한다.

물론 2^1 위치에는 2V, 2^2 위치에는 4V가 되어야 한다.

따라서 최하위 자리값(LSB)의 가중치는 다음 식에 의해 구할 수가 있다.

$$LSB \text{ 가중치} = \frac{1}{2^n - 1}; \text{ 단 } n : \text{비트수}$$

이때 2^n을 **D/A 변환기의 출력 레벨 수**라고 하고, $2^n - 1$을 **스텝**(Step) 수라고 한다.

(2) 하중(Weight) 저항을 이용한 병렬형 D/A 변환기

그림 13 - 2는 병렬형 D/A 변환기의 기본 원리도이다. 이 그림에서 스위치가 ON 되었을 때를 논리 '1', OFF 되었을 때를 논리 '0'이라 하면 그림과 같이 스위칭이 되어 있는 경우('101') 전류 I는 다음과 같이 된다.

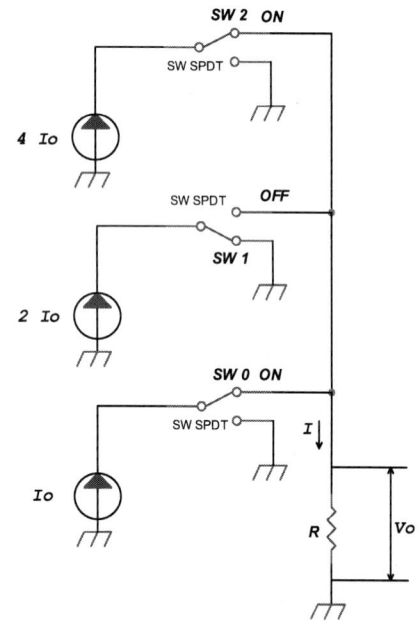

$I = 4I_0 + 0I_0 + 1I_0 = 5I_0$가 되어 R 양단의 전압 Vo는

$V_o = RI = 5I_0$가 되므로 디지털 입력값 '101'에 해당하는 아날로그 출력전압을 얻을 수가 있다.

그림 13 - 2 병렬형 D/A 변환기의 원리

실제 회로에서는 디지털 신호의 출력이 정전류(定電流)로 되는 경우는 거의 없고 대부분이 전압원이므로 그림 13-3과 같이 하중 저항을 통해 등가 전류원으로 변환시켜 주어야 한다. 이 그림과 같이 레지스터에 기억되어 있는 디지털값을 아날로그 신호로 바꾸는 경우를 생각해 보자.

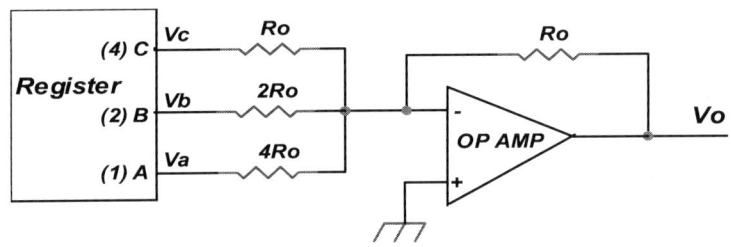

그림 13-3 하중 저항을 이용한 병렬형 D/A 변환기

출력전압 Vo는

$$V_o = -\left(\frac{R_o}{R_o}V_c + \frac{R_o}{2R_o}V_b + \frac{R_o}{4R_o}V_a\right)$$
$$= -\left(V_c + \frac{1}{2}V_b + \frac{1}{4}V_a\right)$$

이 되고 Va~Vc는 레지스터가 이상적인 TTL IC인 경우 '0' 또는 '+5V'의 두 가지 경우뿐이므로 그림 13-2와 같이 스위치가 ON, OFF 된 상태로 대응되게 할 수 있으므로 레지스터에 축적된 디지털 신호값을 이에 대응되게 아날로그 출력전압으로 바꿀 수가 있게 되는 것이다.

위의 그림에서처럼 D/A 변환을 하는 데는 디지털 정보를 저장하기 위한 **레지스터**, 레지스터의 입력을 게이팅(Gating)시킬 수 있는 **게이트, 저항형 회로망, 레벨 증폭기(OP AMP)** 등이 있어야 한다. 레지스터는 보통 매 비트당 하나의 플립플롭을 가진 RS 플립플롭으로 구성된다. 그림 13-4에 이의 개요도가, 그림 13-5에 4 비트 D/A 변환기 구성도가 있다.

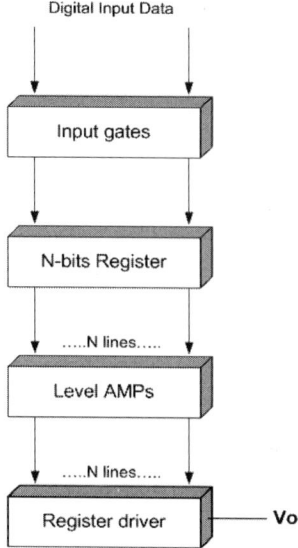

Digital Input Data

Input gates

N-bits Register

.....N lines.....

Level AMPs

.....N lines.....

Register driver — **Vo**

그림 13 - 4 D/A 변환기 개요도

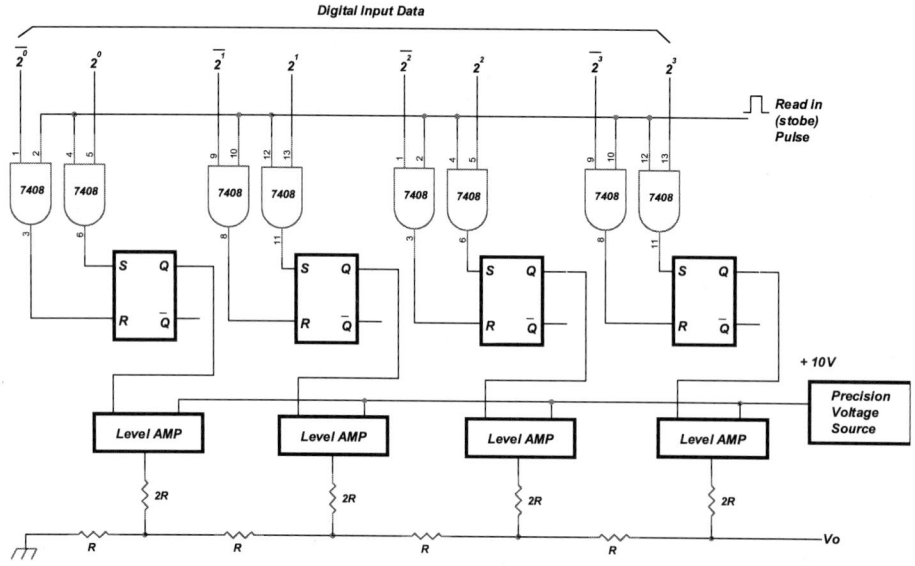

그림 13 - 5 4비트 D/A 변환기 구성도

〈그림 13-5 구성도 동작 설명〉

그림 13-5를 보면 그 모양이 사다리꼴인 것을 알 수 있다.

❶ 각 레벨 증폭기는 2개의 입력 단자를 가지고 있는데 하나는 정밀한 전압원으로부터 오는 +10V이고 다른 하나는 플립플롭으로부터 온다.

❷ 이 증폭기는 플립플롭에서 들어오는 입력이 High 전압 상태일 때 증폭기의 출력이 +10V가 되도록 동작하고 Low 전압 상태이면 출력이 0V가 된다.

❸ 4개의 RS 플립플롭은 디지털 정보를 저장하는 데 필요한 레지스터가 된다. 가장 오른쪽의 플립플롭이 MSB를, 가장 왼쪽의 플립플롭이 LSB를 나타낸다.

❹ 게이트의 입력에 연결된 Read in 선이 High가 되면 각 플립플롭에 연결된 2개의 게이트 중에 하나는 High가 되며 플립플롭은 셋 또는 리셋된다.

❺ 따라서 데이터는 Read in 선에서 펄스가 발생할 때마다 레지스터로 읽히(Read in)게 된다.

❀ 집적회로 D/A 변환기 ❀

D/A 변환기는 쉽게 구할 수 있는 IC이다. 변환기는 6, 8, 10, 12비트의 분해능을 가진 것이 있다. 이 중 대표적인 것이 National semiconductor사의 8비트 D/A 변환기인 DAC 0808이다. 여기서 간단히 DAC 0808의 동작 특성을 설명하겠다. DAC 0808은 8개의 디지털 입력 모두가 '1'일 때 $V_o = +10 V_{dc}$를 발생하도록 되어 있다. 그림 13-6에 이를 이용한 회로도가 있다.

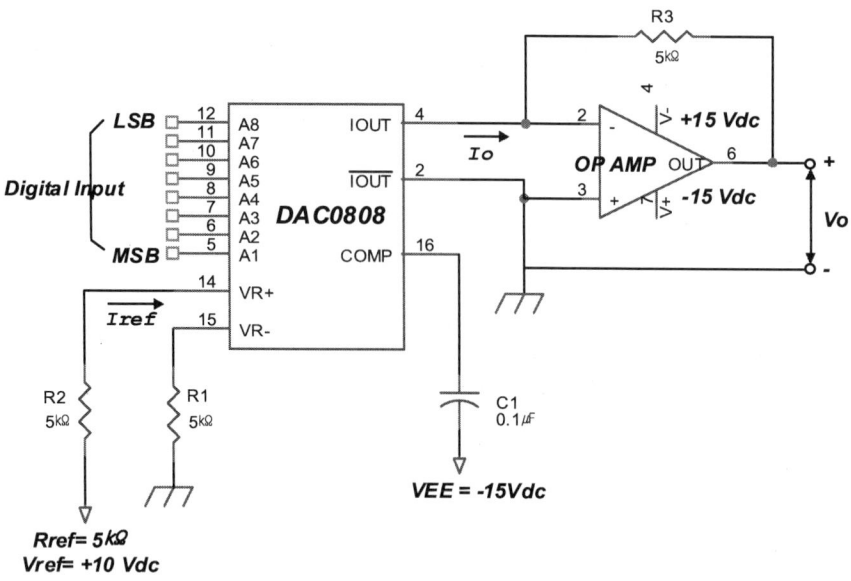

그림 13 - 6 D/A 변환기 사용례(DAC 0808)

〈동작 설명〉

❶ DAC 0808에는 두 개의 전압원($V_{CC} = +5\,V_{dc}$, $V_{EE} = -15\,V_{dc}$)이 필요하다.

❷ V_{EE}에 연결된 $0.1\mu F$의 콘덴서는 오실레이션 현상을 방지하고, V_{EE}에 영향을 받지 않도록 하기 위한 것임.

❸ 4번 핀에는 D/A 변환기의 출력 단자를 연결하는데 아주 제한된 전압 범위($+0.5V \sim -0.6V$)를 갖는다. 이것은 출력전류 I_o를 공급한다. 최소 전류값은 0mA 이고, 최대 전류값은 기준전류 I_{ref} 값과 같다.

〈출력전압의 계산〉

❶ 기준전류 I_{ref}는 다음 식에 의해 구해진다.

$$I_{ref} = \frac{V_{ref}}{R_{ref}}$$

❷ D/A 변환기의 출력전류 I_o는

$$I_o = I_{ref}\left(\frac{A_1}{2} + \frac{A_2}{4} + \frac{A_3}{8} + \cdot \cdot \cdot + \frac{A_8}{256}\right)$$

단, $A_1, A_2, A_3, \cdots\cdots A_8$: 디지털 입력 레벨

❸ OP AMP의 출력전압 V_o는

$$V_o = I_o \cdot R$$

❹ 위 식들에 의해

$$1\,V_o = \frac{V_{ref}}{R_{ref}} \cdot \left(\frac{A_1}{2} + \frac{A_2}{4} + \frac{A_3}{8} + \cdot \cdot \cdot + \frac{A_8}{256}\right) \times R$$

❺ 여기서 OP AMP의 피드백 저항 R이 R_{ref}와 같다면

$$V_o = V_{ref} \cdot \left(\frac{A_1}{2} + \frac{A_2}{4} + \frac{A_3}{8} + \cdot \cdot \cdot + \frac{A_8}{256}\right)$$

❻ 모든 디지털 입력($A_1, A_2, A_3, \cdots\cdots A_8$)이 '0'이라면

$$V_o = V_{ref} \cdot \left(\frac{0}{2} + \frac{0}{4} + \frac{0}{8} + \cdot \cdot \cdot + \frac{0}{256}\right) = V_{ef} \times 0 = 0\,V_{dc}$$

❼ 모든 디지털 입력이 '1'이라면

$$V_o = V_{ref} \cdot \left(\frac{1}{2} + \frac{1}{4} + \frac{1}{8} + \cdot \cdot \cdot + \frac{1}{256}\right) = V_{ref} \times \frac{255}{256} = 0.996 \times V_{ref}$$

이 되고 V_{ref}가 +10V인 경우 출력전압은 0~+9.96V 사이에서 나타나게 된다.

2. 연산증폭기(OP AMP: Operational Amplifier)

(1) 개요

앞에서 살펴본 바와 같이 D/A 및 A/D 변환을 하기 위해서는 연산증폭기의 동작 특성을 알아야 할 필요성이 있다. 따라서 여기서는 연산증폭기의 동작 특성에 대해서 간단히 살펴보기로 하자.

연산증폭기는 초기에는 아날로그 계산기 등에 많이 사용되었으나 최근에는 능동 필터(Active filter), 미적분기(Differentiator, Integrator), 비교기(Comparator), 신호변환기, 함수발생기, 서보 모터 제어, 각종 통신기기 등 아주 많은 부분에서 사용되고 있다.

(2) 연산증폭기의 특성

일반적으로 연산증폭기는 이상증폭기(理想增幅器)의 개념으로 해석한다. 이상증폭기란 다음의 특성을 가진 증폭기를 말한다.

❶ 전압 이득(Av) = 무한대
❷ 입력 임피던스(Zin) = 무한대
❸ 출력 임피던스(Zout) = 0(Zero)
❹ 잡음 = 0(Zero)
❺ 입력 바이어스 전류 = 0(Zero)
❻ 오프셋(Offset) 전류 = 0(Zero)
❼ 오프셋(Offset) 전압 = 0(Zero)
❽ 오프셋 전압 및 전류의 온도 변화와 장시간 방치한 경우의 Drift = 0(Zero)
❾ 주파수 대역폭은 직류에서부터 무한대까지

실제로는 위와 같은 조건을 모두 만족할 수는 없고 가급적이면 이와 유사하게 접근토록 추구하고 있다.

그림 13 - 7은 이상증폭기로서의 OP AMP의 등가회로이다.

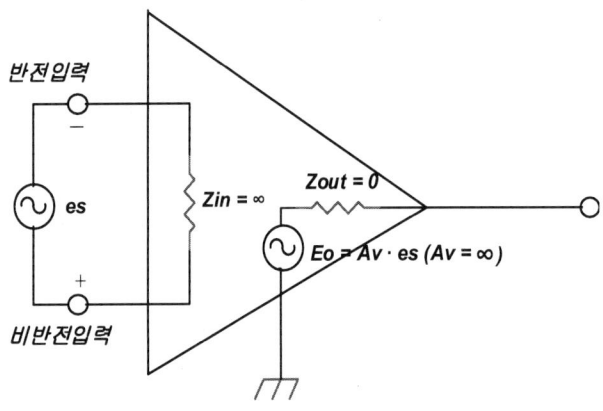

그림 13 - 7 이상적인 연산증폭기(OP AMP)의 등가회로

(3) 입력전류와 출력전압과의 관계

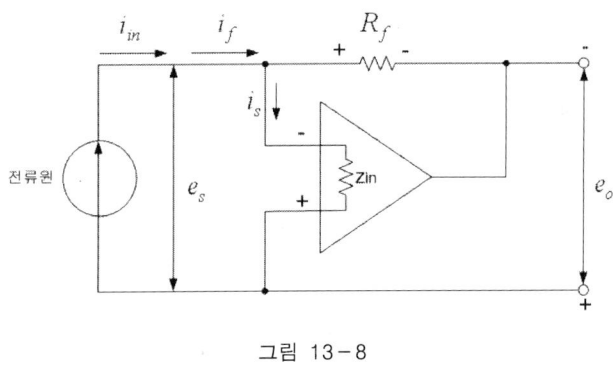

그림 13 - 8

그림 13 - 8과 같이 OP AMP가 반전(Inverting)되어 접속되어 있는 경우 입력전압을 e_s, 출력전압을 e_o, 전압이득을 A라고 하면 출력전압 e_o는

$$e_o = -A e_s \quad \text{...}(13-1)$$

R_f 양단의 전압은 $e_s = -e_o$가 되므로

$$e_s - e_o = e_s + A\,e_s = e_s(1+A) \quad \text{.............. } (13-2)$$

따라서 R_f로 흐르는 전류 i_f는

$$i_f = \frac{e_s - e_o}{R_f} = \frac{e_s(1+A)}{R_f} \quad \text{..................... } (13-3)$$

그런데 입력 임피던스 Z_{in}은 거의 무한대이므로 $i_s \simeq 0$가 되어

$$i_f = i_{in} \quad \text{.................................... } (13-4)$$

따라서 이 경우 입력 임피던스를 Z_s라고 하면

$$Z_s = \frac{e_s}{i_{in}} = \frac{e_s}{i_f} = \frac{R_f}{1+A} \quad \text{............................ } (13-5)$$

가 되며 식 $13-4$와 $13-5$에서

$$e_s = \frac{i_{in}R_f}{1+A} \quad \text{............................... } (13-6)$$

이므로

$$e_o = -A\,e_s = -\frac{A}{1+A}R_f\,i_{in} \quad \text{.................. } (13-7)$$

여기서 $A \gg 1$인 경우라면

$$e_o = -i_{in}R_f \quad \text{................................. } (13-8)$$

가 된다.

즉 OP AMP의 출력전압은 내부 회로적 요소에 관계없이 입력전류(i_{in})와 궤환저
항(R_f)에 의해 결정된다.

(4) 입력전압과 출력전압과의 관계

가. 반전증폭기

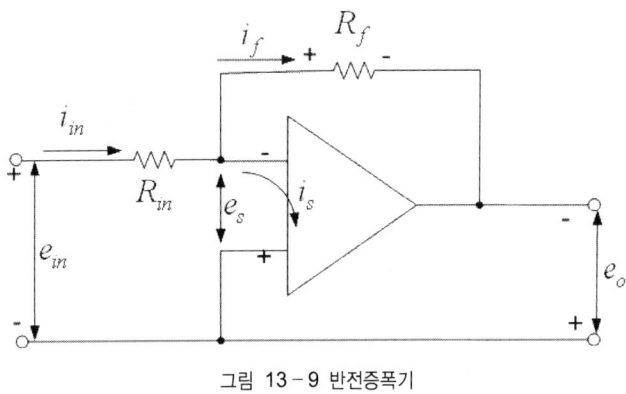

그림 13-9 반전증폭기

그림 13-9의 반전증폭기 회로에서 OP AMP의 입력 임피던스가 거의 무한대라면 $i_s = 0$가 되므로 $i_{in} = i_f$가 되고

$$i_{in} = i_f = \frac{e_{in} - e_s}{R_{in}} = \frac{e_s - e_o}{R_f} \quad\cdots\cdots\cdots\cdots\cdots (13-9)$$

가 된다. 그런데 OP AMP의 증폭도가 매우 크게 되면 $e_s \simeq 0$가 되어야 하므로

$$\frac{e_{in}}{R_{in}} = - \frac{e_o}{R_f} \quad\cdots\cdots\cdots\cdots\cdots\cdots\cdots\cdots (13-10)$$

따라서

$$e_o = - e_{in}(\frac{R_f}{R_{in}}) = - e_{in}A \quad\cdots\cdots\cdots\cdots (13-11)$$

$$\text{단 } A = \frac{R_f}{R_{in}}$$

가 되어 OP AMP와 관계없이 입력저항(R_{in})과 궤환저항(R_f)만으로 이득이 결정되며 식 13 – 11의 앞에 ' – ' 부호가 붙으므로 흔히 이런 회로를 **반전증폭기**라 부른다.

나. 비반전증폭기

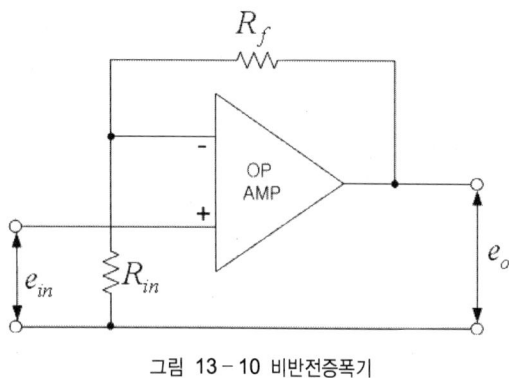

그림 13 – 10 비반전증폭기

위 그림은 비반전증폭기 회로도이다. 여기서 출력전압 e_o는 다음과 같이 주어진다.

$$e_o = (1 + \frac{R_f}{R_{in}})\,e_i = A\,e_i \quad \text{......................} \quad (13-12)$$

$$단\ A = 1 + \frac{R_f}{R_{in}}$$

위 식에서 ' – ' 부호가 없으므로 입력과 출력 간의 위상은 변함이 없다는 것을 알 수 있다.

다. 차동증폭기

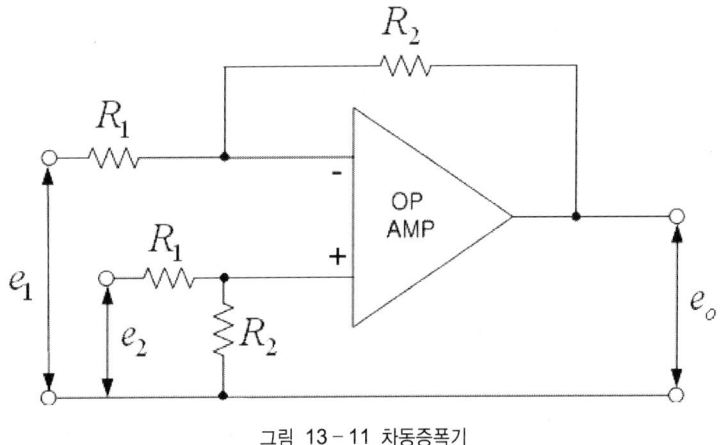

그림 13 - 11 차동증폭기

OP AMP의 초단 증폭회로는 일반적으로 차동증폭기로 되어 있기 때문에 반전증폭기와 비반전증폭기를 그림 13 - 11과 같이 동시에 이용하면 차동증폭기를 구성할 수 있다.

또한 이 회로에서 출력전압 e_o는 다음과 같다.

$$e_o = -\frac{R_2}{R_1}(e_1 - e_2)$$ .... (13 - 13)

(5) OP AMP를 이용한 연산회로

가. 가산기(Adder)

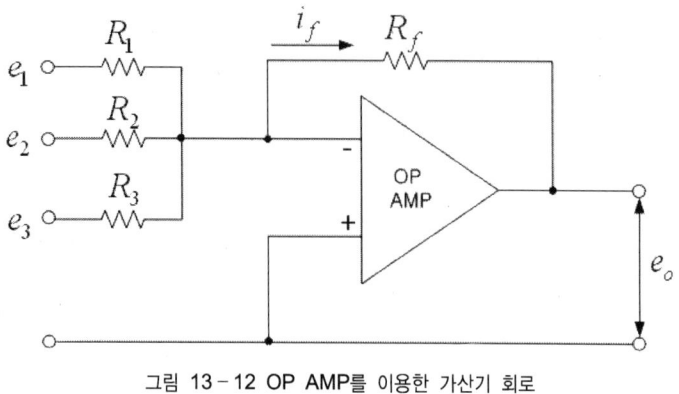

그림 13 - 12 OP AMP를 이용한 가산기 회로

위 그림에서 전류 i_f는

$$i_f = \frac{e_1}{R_1} + \frac{e_2}{R_2} + \frac{e_3}{R_3} \quad \text{..................................} \quad (13-14)$$

이 되고 출력전압은

$$e_o = -\,i_f R_i = -\left(\frac{e_1}{R_1} + \frac{e_2}{R_2} + \frac{e_3}{R_3}\right) R_f \quad \text{...} \quad (13-15)$$

여기서 $R_1 = R_2 = R_3 = R$이면 출력전압은 아래와 같다.

$$e_o = -\left(e_1 + e_2 + e_3\right) \frac{R_f}{R} \quad \text{..........................} \quad (13-16)$$

이 되어 가산기로서 동작을 하게 된다.

나. 적분기(Integrator)

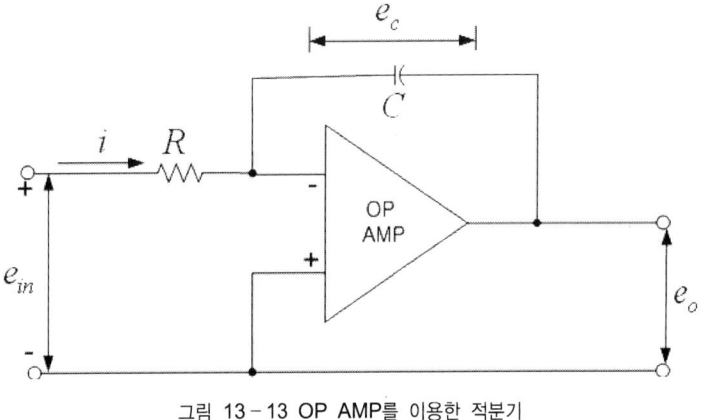

그림 13 - 13 OP AMP를 이용한 적분기

그림 13 - 13에서 입력전류 i는 아래와 같다.

$$i = \frac{e_{in}}{R} \quad \text{...} \quad (13-17)$$

C 양단 전압 e_c는

$$e_c = \frac{1}{C} \int i \, dt = \frac{1}{RC} \int e_{in} dt \quad \text{..................} \quad (13-18)$$

그런데 $e_o = -e_c$가 되어야 하므로

$$e_o = -RC \frac{de_{in}}{dt} \quad \text{.....................................} \quad (13-19)$$

가 되어 출력전압(e_o)는 입력전압(e_{in})을 적분한 상태가 되므로 이와 같은 회로를 적분기라고 한다.

다. 미분기(Differentiator)

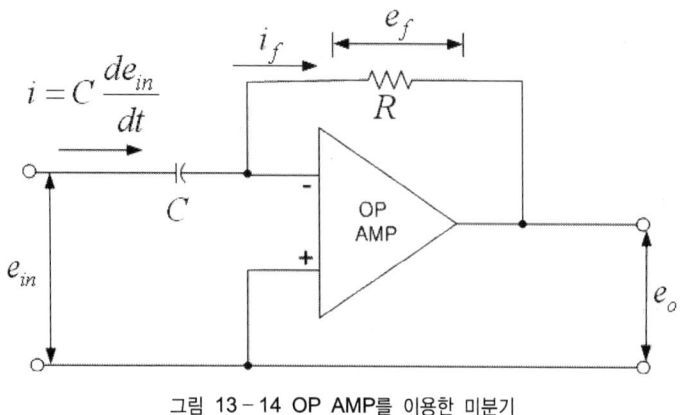

그림 13 - 14 OP AMP를 이용한 미분기

그림 13 - 14에서 출력전압(e_o)는

$$e_o = - RC \frac{de_{in}}{dt}$$ .... (13 - 20)

이 되므로 출력전압(e_o)는 입력전압(e_{in})을 미분한 상태가 된다. 따라서 이와 같은 회로를 미분기라고 한다. 미분회로는 일종의 고역여파기(High pass filter)이므로 OP AMP의 주파수 특성이 우수할수록 고주파 잡음이 잘 증폭되어 전달된다.

(1) 다음 지시에 따라 실험을 진행하시오.

❶ 그림 13 – 15의 회로를 구성하시오.

❷ 회로 구성이 끝나면 회로에서 지시된 전압을 인가하시오.

❸ 16진 카운터 7493을 리셋(reset)시키고 (SW1을 +5V 측으로 한다) 10㏀ 가변 저항을 조정하여 $V_o = 0$가 되게 하시오.

❹ 7493을 셋(set)시키고 (SW1을 접지로) Cp 단자(7493 pin14)에 단일 펄스 발생기를 접속 후 클럭펄스를 1개씩 순차적으로 인가하면서 V_o, V_d, V_c, V_b, V_a 지점의 전압을 측정하여 해당 표에 기록하시오.

❺ Cp 단자에 단일 펄스발생기 대신 1KHz, 5Vp – p 구형파 발진기를 접속하고 Cp와 출력 파형을 비교하여 파형의 모양을 그리시오.

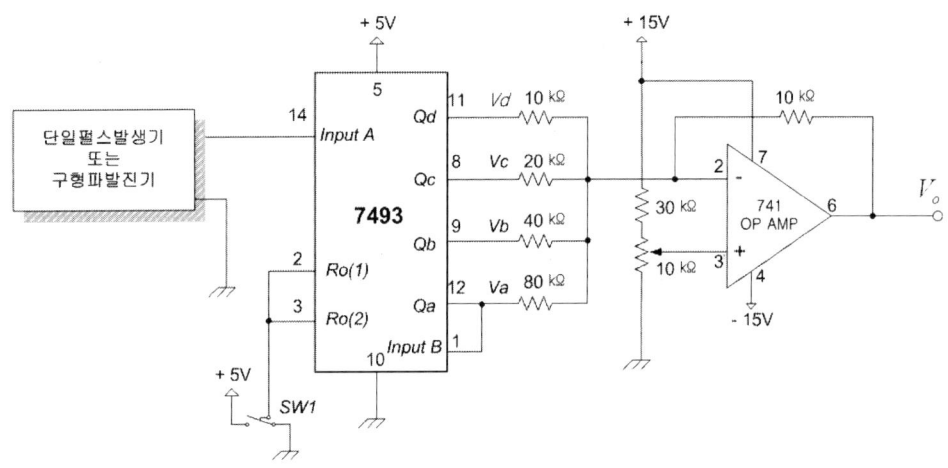

그림 13 – 15

표 13-3

Cp의 인가 수	Vd	Vc	Vb	Va	Vo
0					
1					
2					
3					
4					
5					
6					
7					
8					
9					
10					
11					
12					
13					
14					
15					
16					
17					
18					
19					

표 13-4 Cp 파형

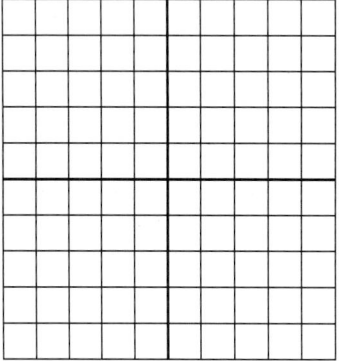

표 13-5 출력 파형

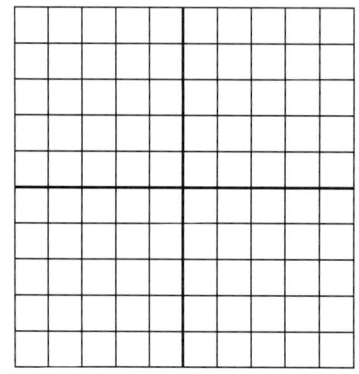

실험결과 Report		학과명	학번	성명
실험 13	D/A 변환			

표 13 - 3

Cp의 인가 수	Vd	Vc	Vb	Va	Vo
0					
1					
2					
3					
4					
5					
6					
7					
8					
9					
10					
11					
12					
13					
14					
15					
16					
17					
18					
19					

표 13 - 4 Cp 파형

표 13 - 5 출력 파형

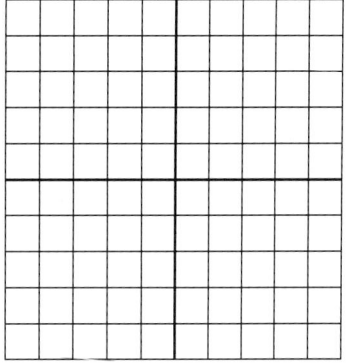

실험 14 — A/D 변환

실험목적

- A/D 변환기의 기본 원리와 의미 및 필요성을 이해하도록 한다.
- 각종 A/D 변환방식의 원리 및 특성을 이해하도록 한다.
- A/D 변환기의 구성방법을 익히고 실험할 수 있도록 한다.

실험기기 및 재료

구분	품명	규격	수량	비고
기기	논리회로 실험장치		1	
	회로시험기		1	
	오실로스코프		1	
	직류전원공급장치		1	±15V, +5V
재료	NAND 게이트	MC 14011	1	CMOS
	BCD Up/Down Counter	MC 14510	1	CMOS
	OP AMP	CA 3160	1	
	LED		1	
	저항	680Ω	1	
	저항	100kΩ	6	
	저항	200kΩ	5	
	가변저항 VR	500kΩ	1	
	점퍼선		약간	

1. A/D 변환(Analog to Digital converter)

(1) 개요

실제 우리 주위에서 얻는 거의 모든 데이터(소리, 온도, 압력 등)는 아날로그 형태로 나타난다. 이러한 아날로그 신호를 디지털 시스템에서 처리하기 위해서는 디지털 신호로 변환해야 하는데 이런 목적으로 사용되는 장치가 아날로그 – 디지털 변환기(Analog to Digital Converter: A/D 변환기 또는 ADC)이다.

이 A/D 변환기는 D/A 변환기와 더불어 자동제어, 음성인식, 영상인식, 의공학 등 많은 부분에서 사용되고 있다.

(2) A/D 변환의 원리

A/D 변환의 원리는 아래 그림 14 – 1에 나타나 있다.

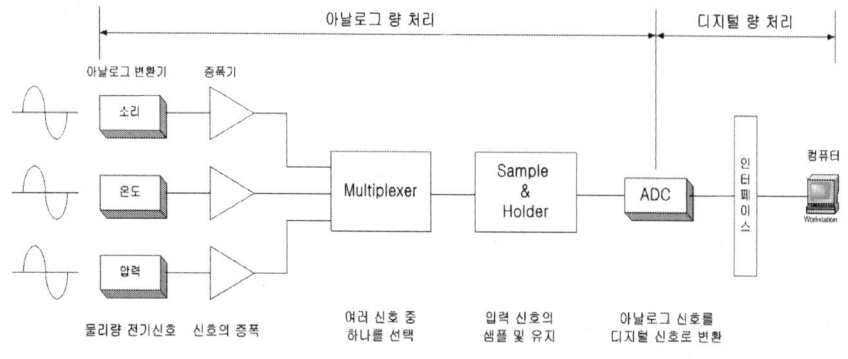

그림 14 – 1 A/D 변환 시스템의 원리

실제 온도, 소리, 압력과 같은 신호는 증폭, 샘플 앤 홀드(Sample & Hold) 등의 과정을 거쳐 A/D 변환기로 입력된다.

(3) 입출력 관계

A/D 변환은 본질적으로 정량화 동작이다. 그림 14 – 2를 보자.

이 그림에서 아날로그 입력신호 V_i는 기준신호 V_{ref}와 비교한 분수값 x로 변환된다. 변환기의 디지털 출력은 이 분수를 코드(Code)로 표현한 것이다. 아래 그림은 이 기본적인 원리를 잘 설명해 주고 있다.

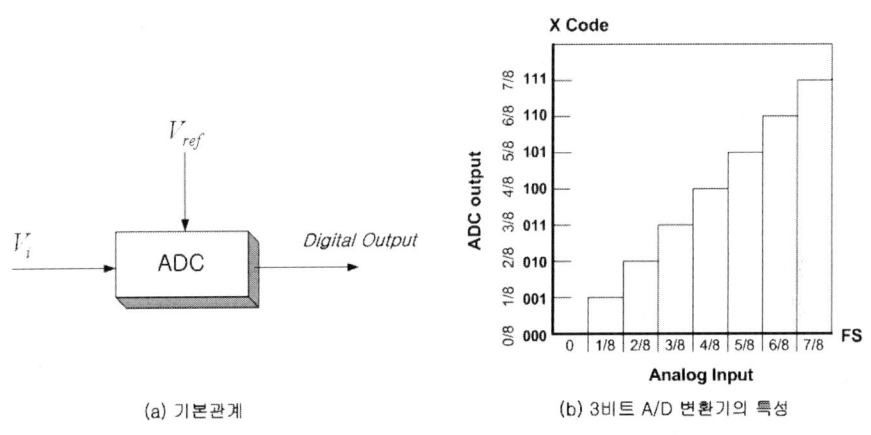

그림 14 – 2 A/D 변환기

변환기의 출력 코드가 n비트로 되어 있으면 이산적 출력등급의 수는 2n으로 된다. 입력 범위도 이와 대응되는 같은 수로 되어야 한다. 등급의 크기는 2개의 인접한 코드의 차에 해당하는 아날로그값이 된다. 이것이 최하위 비트(LSB)의 크기이다. 따라서 LSB의 크기는 다음 식에 의해 구할 수가 있다.

$$Q(LSB) = \frac{FS}{2^n}$$

Q: 한 등급의 크기

LSB: LSB 아날로그값

FS: Full Scale 아날로그 입력크기

앞의 그림은 이상적인 3비트 ADC에 대한 변환관계를 나타내고 있다. LSB의 크기는 1/8 *FS*이고 입력 범위는 0 ~ 7/8 *FS*까지 8개의 구별이 가능한 크기로 양자화된다. 여기서 2진수 '111'의 최대 출력은 Full scale에 대응되는 것이 아니라 7/8 *FS*에 대응되고 있는 것에 주목하라. 0점이 코드 '000'을 가지므로 ADC의 출력은 항상 *FS*에서 1 LSB를 뺀 아날로그값에 대응된다(0 ~ 7/8 *FS*).

한 등급 내의 모든 아날로그값들은 Threshold라 불리는 중간값에 대응하는 디지털 코드로 나타낸다. 입력신호는 ±1/2 LSB만큼 Threshold값과 차이가 날 수 있고 같은 출력 코드로 나타내지므로 A/D 변환 과정에서는 ±1/2 LSB의 양자화 잡음이 존재하게 된다. 즉 1.5/8 *FS*의 값은 디지털 코드로 '000'과 '001' 사이의 값이 되어야 하는데 디지털값은 '000'과 '001' 사이 값이 존재하지 않는다. 따라서 '000' 또는 '001'로 양자화되어야 한다. 여기에서 디지털 코드로 양자화된 값은 원래의 값과 0.5만큼의 오차가 발생하게 되는 것이다. 양자화 잡음이란 바로 이러한 오차가 생기는 현상을 말한다. 이 잡음은 변환기의 출력 코드에서 비트의 수를 증가시킴으로써 최소화할 수 있다.

(4) A/D 변환방식

A/D 변환은 D/A 변환에 비해 복잡하고 더 많은 시간이 소요되며 A/D 변환에는 여러 가지 방법들이 있다. 다음에 A/D 변환을 위한 몇 가지 방법들에 대해 살펴보기로 하자.

(가) 병렬비교형 A/D 변환방식

이 방식은 영상, 레이더, 디지털 오실로스코프 등과 같이 빠른 속도의 변환이 요구되는 시스템에 주로 사용되는 방식으로 기본 회로의 구성은 아래와 같다. 이 변환기는 **동시형 A/D 변환기**(Simultaneous A/D Converter)라고도 한다.

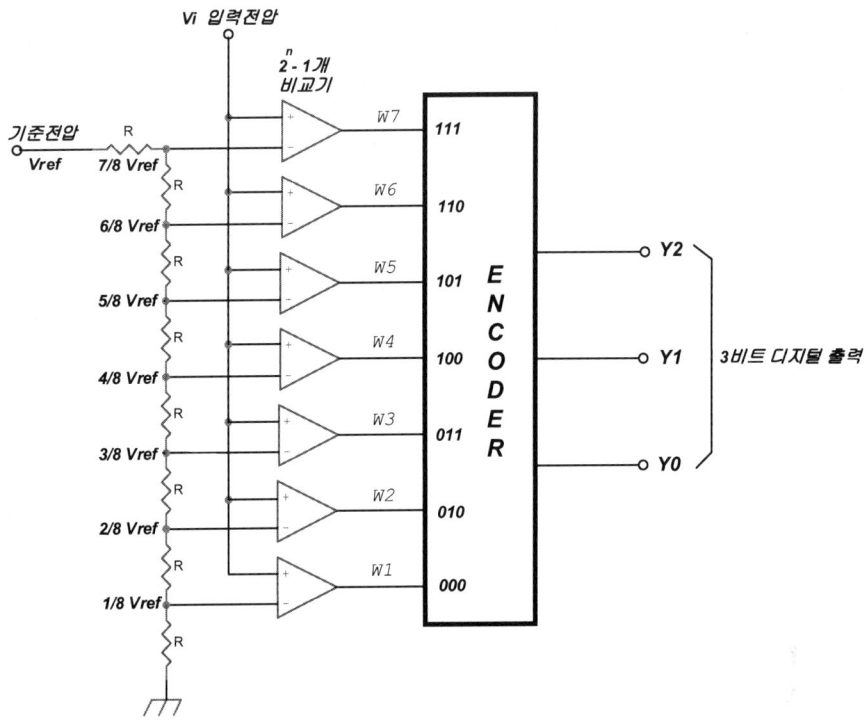

그림 14 – 3 3비트 병렬비교형 A/D 변환기

이 회로에서 비교기는 n비트 A/D 변환기인 경우 2n – 1개가 필요하게 된다. 즉 3비트 병렬비교형 A/D 변환기의 경우 23 – 1＝7개의 비교기가 필요하게 된다. 각 비교기는 아날로그 입력신호와 기준전압을 저항들로 분할한 전압과 비교하여 'H' 와 'L'의 디지털값을 출력한다. 이러한 비교기의 출력은 인코더 회로에 의해 2진 코드로 변환된다.

위의 회로에서
입력전압 V_i가 7.5V이고, 기준전압 V_{ref}가 20V일 때

$$V_i = \frac{x}{8} \times V_{ref}$$
$$7.5 = \frac{x}{8} \times 20$$

에서 x의 값이 3이 되므로 비교기의 출력은 W_1, W_2, W_3에서 나오게 된다. 따라서 Encoder는 이 입력을 받아 3비트 '011'의 디지털 코드를 출력하게 되는 것이다. 다음 표 14 – 1에 3비트 병렬비교형 A/D 변환기의 진리표가 있다.

표 14 – 1 3비트 병렬비교형 A/D 변환기

입력							출력		
W7	W6	W5	W4	W3	W2	W1	Y2	Y1	Y0
0	0	0	0	0	0	0	0	0	0
0	0	0	0	0	0	1	0	0	1
0	0	0	0	0	1	1	0	1	0
0	0	0	0	1	1	1	0	1	1
0	0	0	1	1	1	1	1	0	0
0	0	1	1	1	1	1	1	0	1
0	1	1	1	1	1	1	1	1	0
1	1	1	1	1	1	1	1	1	1

✍ 병렬비교형 A/D 변환기는 다음과 같은 특징을 가지고 있다.

❶ A/D 변환속도가 비트 수에 관계없이 일정하고 매우 빠르다.

❷ 비트 수가 많아질수록 회로가 복잡하게 된다.

　(분해능이 n비트 A/D 변환기의 경우 2n – 1개의 비교기가 필요하므로)

❸ 따라서 비트 수를 늘릴 경우 가격이 비싸진다.

(나) 축차비교형 A/D 변환방식

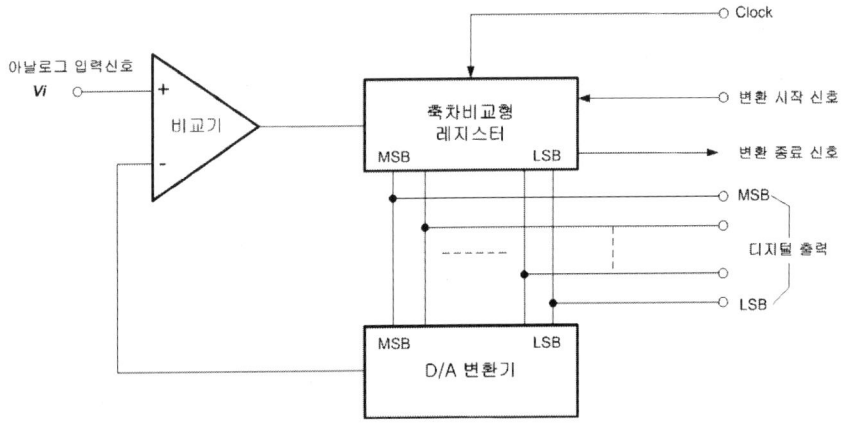

그림 14 - 4 축차비교형 A/D 변환기

축차비교형 A/D 변환기의 기본 회로는 위 그림과 같다. 이 변환기는 연속근사 A/D 변환기라고도 하며 가장 널리 사용되는 변환기 중의 하나로서 아날로그 입력에 관계없이 고정된 변환 시간을 갖는다. 구성 형태는 비교기, D/A 변환기, 축차비교형 레지스터(Successive approximation register)로 구성되어 있다.

이 방식에서는 디지털 출력 비트와 동일한 비트 수의 D/A 변환기를 사용하며, 그 출력이 아날로그 입력전압에 알맞은 근삿값으로 되는 D/A 변환기의 입력 디지털값을 결정함으로써 A/D 변환의 결과를 얻는다.

축차비교형 A/D 변환기에서 아날로그 입력신호는 다음과 같은 순서에 의해 디지털값으로 변환된다.

⟨동작 과정⟩
❶ 변환 시작 신호를 입력한다.
❷ 축차비교 레지스터의 최상위 비트(MSB)를 '1'로 하여 D/A 변환기의 출력전압이 전체 측정 범위의 1/2이 되도록 한다.
❸ 비교기는 D/A 변환기의 출력전압과 입력신호전압을 서로 비교한다.

❹ 입력신호전압이 D/A 변환기의 출력전압보다 크면 축차비교 레지스터의 최상위 비트(MSB)를 '1'로 하고, 작으면 '0'으로 한다.

❺ MSB는 그 상태로 유지하고 다음 비트를 '1'로 하고 D/A 변환기의 출력전압을 다시 비교해서 입력신호가 D/A 변환기의 출력보다 크면 '1'로 하고, 작으면 '0'으로 한다.

❻ 같은 방법으로 LSB까지 행하면 D/A 변환기의 출력은 그림 14-5처럼 점차 아날로그 신호에 접근하게 된다. 이 그림은 6비트 A/D 변환기로 변환결과가 011010의 값을 가진 경우를 보여주고 있다.

축차비교형 A/D 변환기는 다음과 같은 특징을 가진다.

❶ 회로 구성이 간단하여 IC화하기가 쉽다.
❷ 변환속도는 디지털 출력 코드의 비트 수에 의해 결정된다.
❸ A/D 변환의 타이밍(Timing)을 외부 클럭과 동기시킬 수 있다.

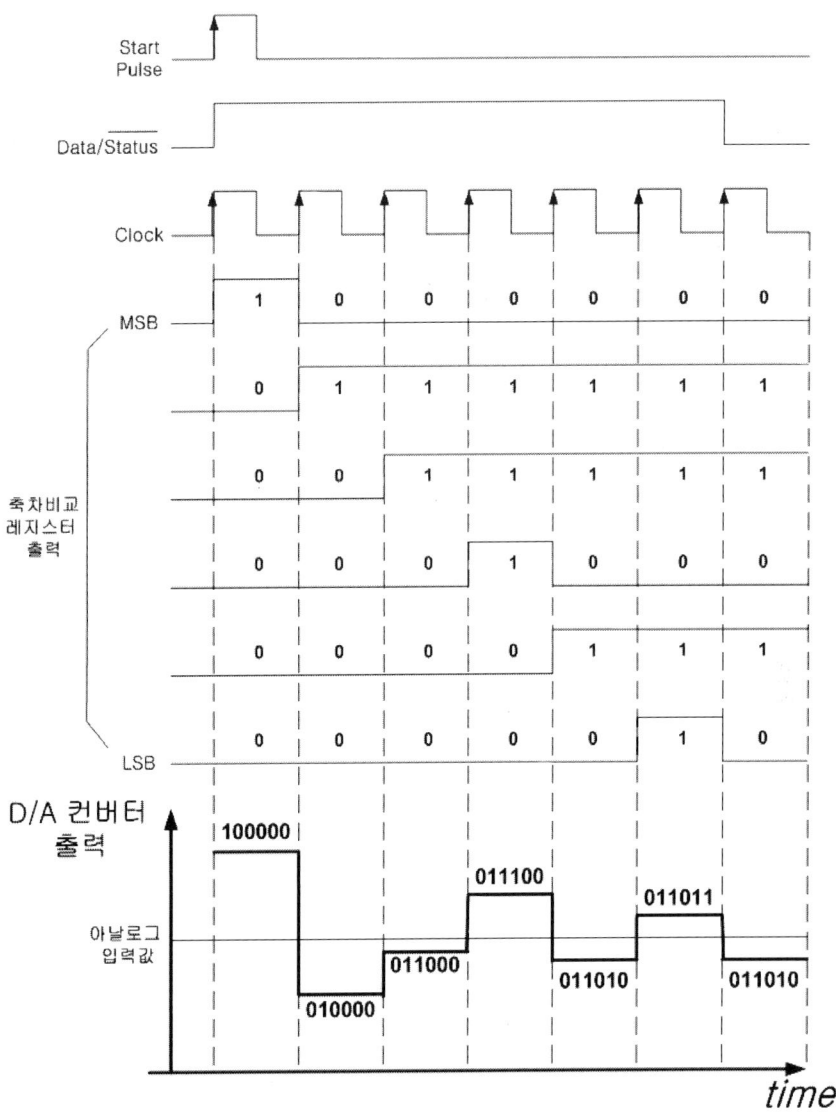

그림 14 - 5 축차비교형 A/D 변환기의 타이밍도

(다) 적분형 A/D 변환방식

이 변환기는 변환속도가 다소 느리지만(10～100ms 정도) 비교적 간단하게 고정밀도의 변환결과를 얻을 수 있는 변환기이다. 이 변환기는 다음 그림과 같이 입력전환 스위치, 적분기, 비교기, 카운터 및 제어 로직(Control logic) 등으로 구성되어 있다.

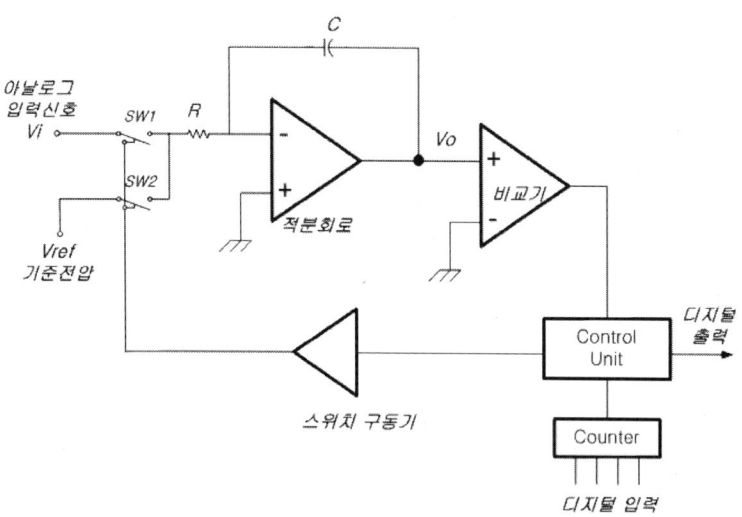

그림 14 - 6 적분형 A/D 변환기

〈동작 과정〉

❶ 컨덴서 C의 전하가 '0'일 때 제어로직에 의해 SW1은 ON, SW2는 OFF 된다.

❷ 일정 시간 T_i 동안 입력전압을 적분한다(입력전압은 일정 시간 콘덴서 C에 충전되며 일정 시간이 흐른 뒤 C의 전압은 입력전압 V_i에 비례하게 된다). 이때 적분 콘덴서 C에 충전되는 전하량 Q_c는 다음과 같다.

$$Q_c = \frac{V_i}{R} T_i$$

V_i : 입력전압$(V), R$: 적분저항$(\Omega), T_i$: 입력신호의 적분시간(sec)

❸ 적분 시간 T_i가 지나면 SW1을 OFF, SW2를 ON으로 하여 입력신호전압의 극성과 반대 극성의 기준전압 V_{ref}로 적분 콘덴서 C의 전하량이 0이 될 때 (T_{ref})까지 방전(역적분)한다. 이때 적분 콘덴서 C에서 방전되는 전하량 Q_D는 다음과 같다.

$$Q_D = \frac{V_{ref}}{R} T_{ref}$$

V_{ref} : 기준전압(V), T_{ref} : 기준전압의 적분시간(sec)

❹ 위 두 식에서 $Q_C = Q_D$이므로 다음의 관계식이 성립하게 된다.

$$V_i = \frac{V_{ref}}{T_i}\ T_{ref}$$

여기서 기준전압 V_{ref}와 입력신호의 적분 시간 T_i는 일정한 값이므로 입력신호 전압 V_i는 역적분 시간 T_{ref}에 비례한다는 것을 알 수 있다. 따라서 역적분 시간 T_{ref}를 정확히 카운팅하면 입력신호 전압의 디지털값을 구할 수 있다.

적분형 A/D 변환기는 다음과 같은 특징을 갖는다.

❶ 적분기에 사용되는 부품의 오차 및 드리프트(Drift)는 변환 정밀도에 영향을 주지 않는다.

❷ 입력 적분 시간의 선택에 의해 입력신호에 포함된 잡음을 필터 없이도 제거할 수 있다(적분회로를 사용하므로 잡음에 면역성이 있다).

❸ 높은 분해능을 비교적 간단히 실현할 수 있다.

❹ 소형화 및 저 가격화가 가능하다.

(1) 다음 지시에 따라 실험을 진행하시오.

❶ 그림 14 - 7의 A/D 변환기 회로를 구성하시오.

❷ 스위치를 누르면서 측정점 A, B, C, D의 전압을 측정하여 표 14 - 2에 기록하시오.

표 14 - 2

입력전압	2진 출력값			
	A	B	C	D
0 V				
0.2 V				
0.4 V				
0.6 V				
0.8 V				
1.0 V				
1.2 V				
1.4 V				
1.6 V				
1.8 V				
2.0 V				
2.2 V				
2.4 V				
2.6 V				
2.8 V				
3.0 V				

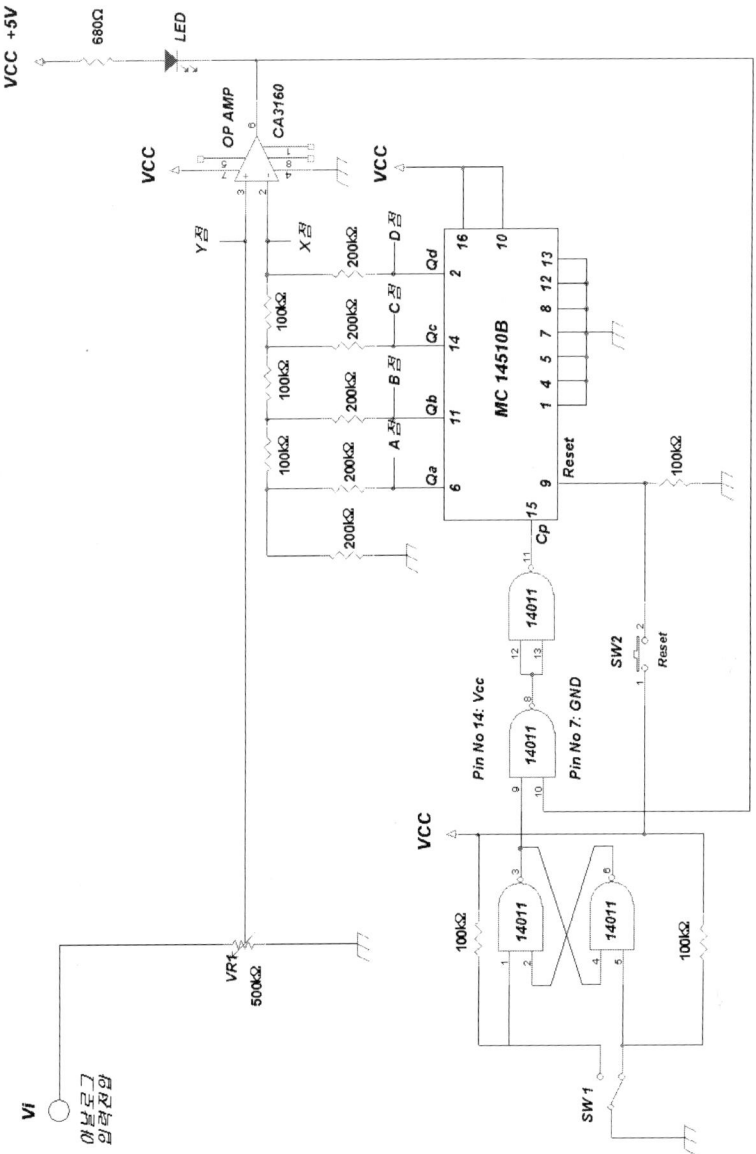

그림 14 – 7

표 14-2

입력전압	2진 출력값			
	A	B	C	D
0 V				
0.2 V				
0.4 V				
0.6 V				
0.8 V				
1.0 V				
1.2 V				
1.4 V				
1.6 V				
1.8 V				
2.0 V				
2.2 V				
2.4 V				
2.6 V				
2.8 V				
3.0 V				

실험 15 LED와 카운터를 이용한 스톱워치

실험목적

- LED(Light emitted diode) 표시기의 원리를 이해하고, 응용할 수 있도록 한다.
- LCD의 원리를 이해하고 디코더를 이용한 LCD 구동방법을 익힌다.
- LED와 카운터를 이용한 스톱워치를 제작하고 그 동작원리를 이해한다.

실험기기 및 재료

구분	품명	규격	수량	비고
기기	논리회로 실험장치		1	
	회로시험기		1	
	오실로스코프		1	
	직류전원공급장치		1	±15V, +5V
재료	NAND 게이트	IC 74LS00	1	2입력
	Counter	IC 74LS90	4	Decade counter
	Schmitt Trigger	IC 74LS14	1	
	BCD to 7-Segment decoder	IC 74LS47	3	
	JK FF	IC 74LS76	1	
	Monostable Multivibrator	IC 74LS121	1	
	Timer IC	555	1	
	FND	7-Segmeent	3	Common Anode
	Push Button Switch		2	
	저항	300 Ω	21	
	저항	820 Ω	2	
	저항	8.2kΩ, 32kΩ	각 1	
	가변저항 VR	500kΩ	1	
	콘덴서	0.012μF	1	
	콘덴서	0.033μF	1	
	콘덴서	22μF	1	
	점퍼선		약간	

이론적 고찰

1. LED 표시기

(1) 개요

지금까지는 카운터 회로에 대해서 실험을 해 보았으나 카운터가 계수한 결과치는 2진수로 표시가 되므로 우리가 눈으로 판독하기에는 매우 불편하였다. 따라서 사람이 판독하기에 편리하도록 표시방법을 달리해야 할 필요성을 느끼게 된다.

이와 같은 용도로 사용되는 것에는 여러 가지가 있으나 가장 대표적인 것이 표시기이다. 표시기는 발광 다이오드(LED)를 여러 개 모아 문자나 숫자를 표시할 수 있도록 되어 있으며 화합물 반도체에 따라서 비화갈륨(GaAs)과 인화갈륨(GaP)으로 구별할 수 있는데 비화갈륨은 적색발광을 하며 최대 파장이 9,400[Å], 인화갈륨은 녹색 발광을 하며 최대파장이 약 5,600[Å] 정도 된다.

Å (옹그스트롱)	거리 단위로 10^{-10}m를 말함

LED는 순방향 전압 1~2[V]를 공급할 때 5~30[mA] 정도의 전류가 흐르며 이에 따라 정해진 파장의 빛을 발하게 된다.

그림 15 - 1은 LED의 구조를 보여준다. 그림 15 - 2는 LED 표시기 조립회로로 7 - 세그먼트(Segment) 숫자 표시기이다. 숫자 표시기에는 애노드 공통 접지형(Common anode type)과 캐소드 공통 접지형(Common cathode type)이 있다. 이 표시기를 사용할 때는 반드시 전류제한 저항을 연결해야 한다. 저항을 연결하지 않으면 과전류로 인해 LED가 파괴될 수도 있다.

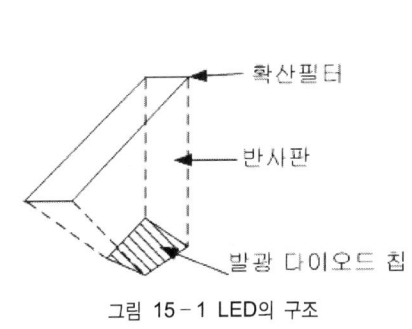

그림 15-1 LED의 구조

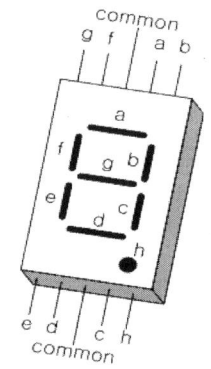

그림 15-2 7-Segment

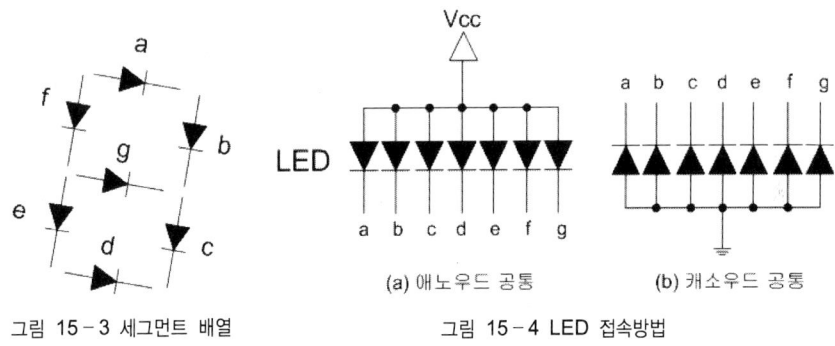

그림 15-3 세그먼트 배열

그림 15-4 LED 접속방법

LED 7-Segment 디코더 드라이버용 IC로는 애노드 공통 접지형의 대표적인 것이 TTL IC의 SN7447이 있고, 캐소드 공통 접지형은 SN7448과 CMOS의 MC14511이 있다. 이들의 사용방법은 뒤에 설명하기로 하겠다.

LED에 접속하여 사용하는 저항값을 어떻게 구하는지 알아보자. 그림 15-5의 회로에서 전압 5V를 사용하고 구동전류는 10mA, LED의 순방향 전압 강하를 2V라면 다음의 공식에 의해 저항값을 구할 수 있다.

$$R = \frac{5-2}{10\text{mA}} = 300\Omega$$

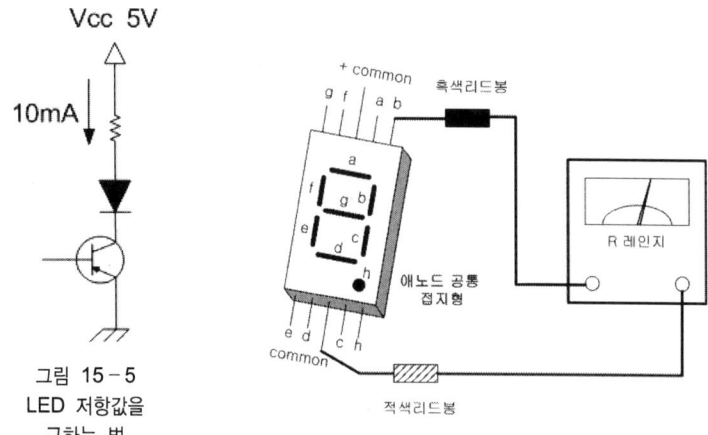

그림 15 - 5
LED 저항값을
구하는 법

그림 15 - 6 LED 표시기(FND 507) 검사방법

LED 표시기의 이상 유무를 점검할 때는 애노드 공통 접지형일 경우 위 그림 15
-6과 같이 회로시험기의 레인지를 R×1에 놓고 FND 507의 공통(Common) 단자에
적색 리드봉을 대고 나머지 단자는 흑색 리드봉을 차례로 대어 a, b, c, d, e, f, g
가 정확히 불이 들어오는지 검사하면 된다. 물론 캐소드 공통 접지형일 경우에는
리드봉을 반대로 연결하면 된다.

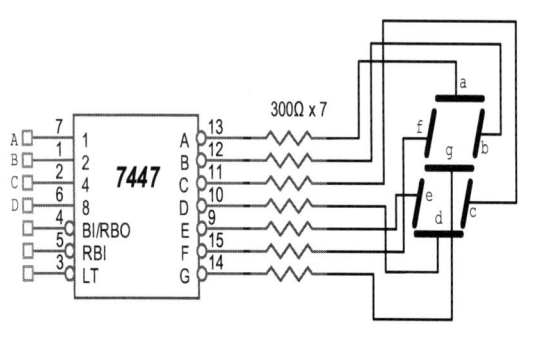

그림 15 - 7 디코더와 7 - Segment의 접속회로

그림 15 - 7은 BCD 숫자를 7
-Segment 형태로 출력해 주는
디코더(7447)와 LED 표시기와의
접속방법을 나타낸다. 주의할 것은
앞에서 이미 설명한 대로 7447
에는 애노드 공통 접지형 LED를
접속하여 사용해야 한다.

표 15 – 1 디코더 입력값에 따른 표시기 표시값

D	C	B	A	표시
0	0	0	0	0
0	0	0	1	1
0	0	1	0	2
0	0	1	1	3
0	1	0	0	4
0	1	0	1	5
0	1	1	0	6
0	1	1	1	7
1	0	0	0	8
1	0	0	1	9
1	0	1	0	A
1	0	1	1	B
1	1	0	0	C
1	1	0	1	D
1	1	1	0	E
1	1	1	1	F

(2) 카운터와 LED 표시기의 접속

카운터와 LED 표시기를 접속하여 사용할 경우에는 카운팅 결과를 LED 표시기에 숫자로 표시해 줄 수 있는 디코더를 반드시 사용해야 한다. 카운터는 2진 상태의 카운팅 결과를 나타내므로 디코더가 없다면 해독하기가 어렵게 된다. 아래 그림 15 – 8에서 이들의 접속방법을 보여준다.

이 회로는 카운터와 디코더 및 LED 표시기로 구성되어 있는데 카운터는 4비트로 된 2진 조합(0000～1111)까지를 출력하고 디코더는 그 값을 받아서 카운터의 출력값이 얼마인지를 해당하는 숫자의 형태로 LED 표시기에 나타내 준다. 단 10(1010)부터 15(1111)까지의 숫자는 A～F의 형태로 표시해 준다.

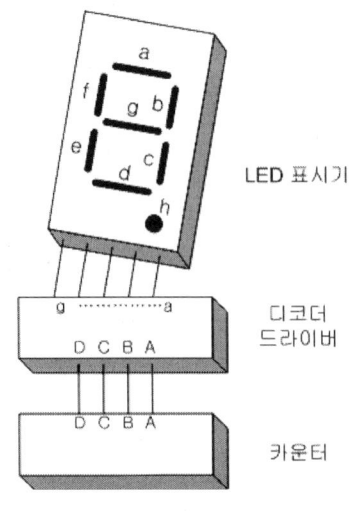

LED 표시기

디코더
드라이버

카운터

그림 15 - 8 카운터와 표시기의 접속

2. 액정 표시기(LCD)

LED 표시기는 전류가 개개의 세그먼트를 통해 흐르면서 빛 에너지를 생성하고
방출되도록 되어 있다. 반면에 액정 표시기는 이용할 수 있는 빛의 반사를 제어한
다. LCD는 다음 그림과 같이 빛을 반사하는 Backplane과 반사된 빛을 보이지 않게
해 주는 세그먼트 부분으로 구성되어 있다.

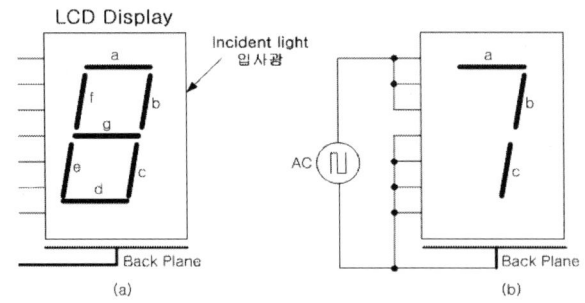

그림 15 - 9 액정 표시기(LCD)

그림 15 - 9(a)는 숫자 판독을 위한 LCD의 세그먼트 배열 상태를 보여주고 있다.
세그먼트를 ON 시키기 위한 AC 전압이 모든 세그먼트에 대해 공통인 세그먼트와

뒤 Backplane에 인가된다. 전형적인 LCD에는 미소한 전류가 흐르고 25~60Hz의 저주파와 낮은 전압(3~15Vrms)에서 동작한다.

LCD에서는 실제로 눈으로 보이는 세그먼트 부분이 빛이 반사되지 않는 부분이 다. Back plane과 세그먼트 간의 전압 차이가 없을 때는 LCD의 세그먼트가 비활성 (OFF) 상태라고 한다. 그림 15-9(b)에서 세그먼트 d, e, f, g는 OFF이고 Backplane에 반해 보이지 않도록 하기 위해 입력된 빛을 반사한다. 적정한 AC 전압이 세그먼트와 Backplane에 인가될 때는 세그먼트가 작동(ON)한다. 그림 15-9(b)의 a, b, c 부분이 바로 빛을 반사하지 않는 부분이 된다. 그래서 이들은 Backplane에 비해 어둡게 보이게 되는 것이다. LCD는 이러한 방법으로 숫자를 표시하기 때문에 어두운 곳에서는 LED에 비해 잘 보이지 않는다는 단점이 있다.

〈LCD의 구동〉

LCD 세그먼트는 세그먼트와 Backplane에 AC 전압이 공급되면 ON 되고 전압이 공급되지 않으면 OFF 된다.

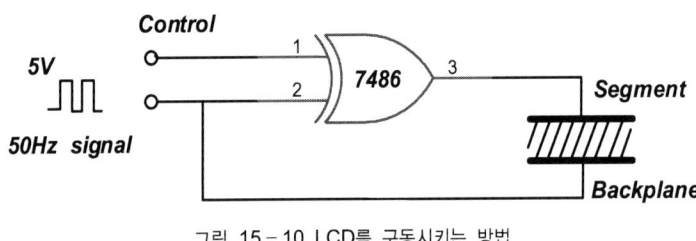

그림 15-10 LCD를 구동시키는 방법

그림 15-10에서 EX-OR의 입력 중 한 단자는 제어 입력단으로 세그먼트의 ON-OFF를 제어하고, 50Hz의 구형파를 다른 한 단자와 Backplane에도 인가한다.

〈동작 과정〉

❶ 제어입력(Control)이 Low일 때

- EX - OR의 출력은 정확히 50Hz이므로, 신호는 세그먼트와 Backplane에 동일
  하게 공급된다.

- 그러면 전압차가 없으므로 세그먼트가 OFF 된다.

❷ 제어입력(Control)이 High일 경우에는

- EX - OR는 50Hz의 역구형파가 출력되므로 세그먼트에 공급되는 신호는 Backplane
  에 인가되는 신호를 초과하게 된다.

- 따라서 세그먼트 전압은 Backplane에 비해 +5V와 -5V로 반복 발생한다. 이
  AC 전압은 세그먼트를 ON 시킨다.

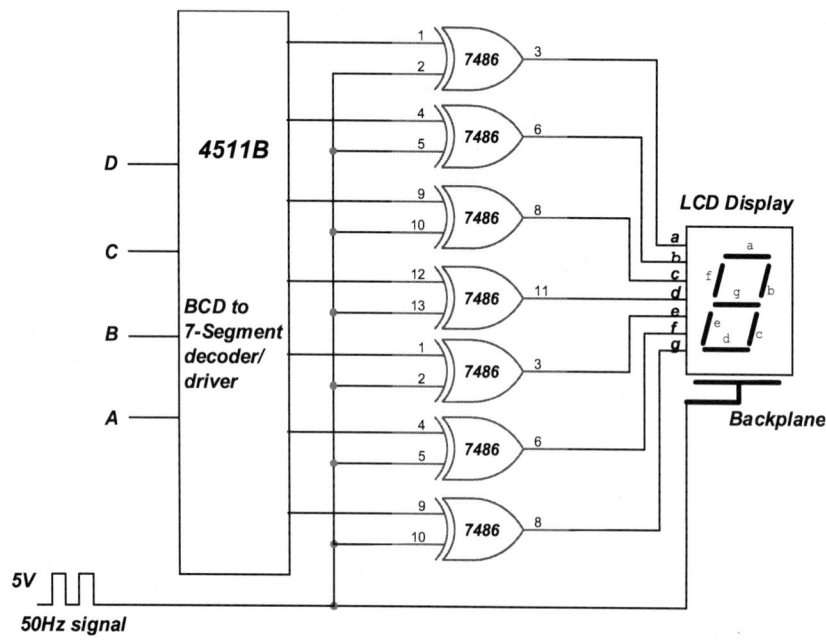

그림 15 - 11 7 - 세그먼트 LCD를 구동시키는 방법

앞에 그림은 CMOS 4511 BCD to 7 – Segment decoder/driver에서 각각 7개의 EX – OR에 제어신호가 공급되는 형태로 되어 있다. 4511은 동적 High를 취하고 있으므로 세그먼트를 ON 시키기 위해 High가 필요하다. 이 그림에서 decoder/driver와 EX – OR이 하나의 칩으로 구현된 것이 있는데 CMOS 74HC4543이 바로 그것이다. 이 IC는 BCD 입력 코드를 갖고 직접 LCD 세그먼트를 구동하기 위한 출력을 제공해 주고 있다.

CMOS 74HC4543과 LCD를 접속한 그림이 아래에 있다.

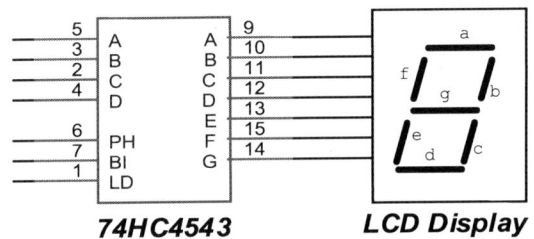

74HC4543　　　　**LCD Display**

그림 15 – 12 74HC4543과 LCD의 연결

일반적으로 LCD를 구동하는 소자로 CMOS를 이용하는데 그 이유는 다음과 같다.

❶ TTL IC보다 소모 전력이 적다.
❷ TTL IC는 LCD의 수명을 단축시킨다.
왜냐하면 TTL IC는 Low일 때 전압은 정확히 0V가 아니라 약 0.4V 정도 된다. 따라서 세그먼트와 Backplane에 DC 성분을 생성하게 되기 때문이다.

3. 스톱워치 제작

본 절에서는 기록경기에서 흔히 사용되는 스톱워치의 원리를 모듈별로 설명하고 0.1초의 정확도를 갖는 스톱워치를 실제 제작하는 방법에 대해 설명하겠다. 스톱워치를 기능별로 나누어 보면 스톱워치에 사용되는 기준 시간을 발생하는 **기준 시각 발생회로**와 시작 시각과 마지막 시각을 결정해 주는 **Start/Stop 회로**, 그리고 0.1초에서 99.9초까지 카운트하고 카운팅 결과를 표시해 주는 **카운터 및 표시회로** 및 **Reset 회로**의 4개 모듈로 구성되어 있다.

이제 각 모듈별 회로와 그 동작특성을 설명하도록 하겠다.

(1) 기준 시각 발생회로

기준 시각 발생회로는 아래 그림과 같이 555 Timer, 슈미트 트리거 IC 7414 및 Decade 카운터 IC7490으로 구성된다. 555 Timer에서 100Hz의 펄스열을 만들어 내고 IC 7414를 이용하여 구형파로 만들고 이를 이용하여 분주회로 IC 7490에서 10Hz의 기준 펄스를 만들어 준다.

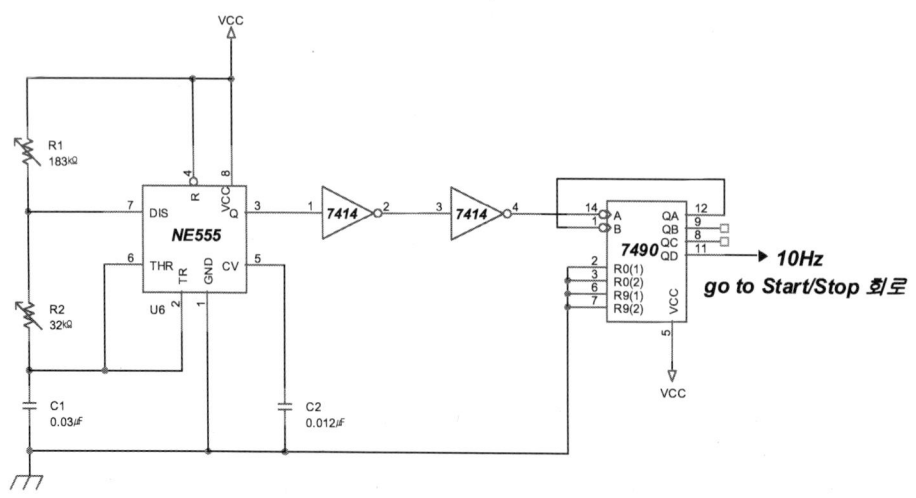

그림 15 - 13 기준 시각 발생회로

(2) Start/Stop 회로

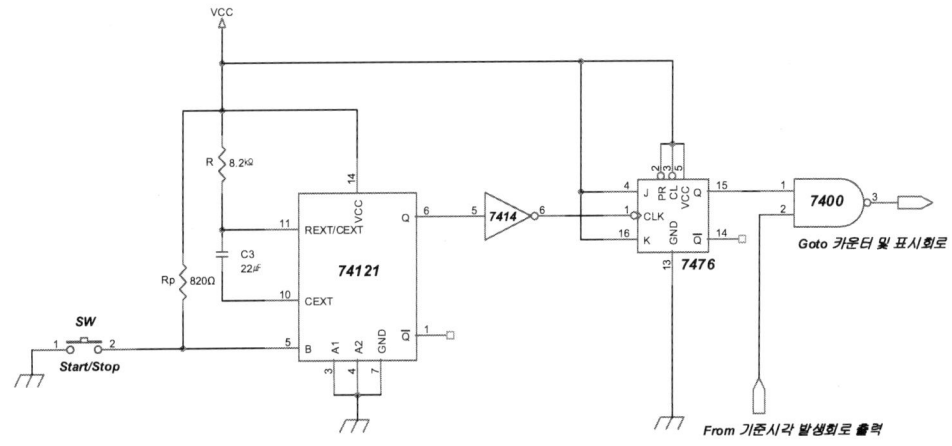

그림 15 - 14 Start/Stop 회로

Start/Stop 회로는 그림 15 - 14와 같이 구성되어 있다. 이 회로는 Start/Stop용 Push - Button 스위치로부터 트리거 펄스를 받아 단안정멀티바이브레이터 IC 74121 을 이용하여 하나의 펄스를 만들고 슈미트트리거를 통해 좀 더 구형파에 가깝게 만든 다음 IC 7476 JK - FF은 클럭펄스로 이용된다. 이 클럭펄스에 의해 IC 7476 의 이전 출력 상태가 반전되어 기준 시각 발생회로로부터 나오는 출력과 NAND되 어 카운터 및 표시회로로 신호가 보내진다. 이때 IC 7476 JK - FF의 출력이 Low가 될 경우 카운터 회로가 동작하지 않아(NAND의 출력이 High) 표시기의 상태는 변 하지 않는다. 한편 IC 7476 JK - FF의 출력이 High가 되면 카운터 회로가 정상적 으로 동작하여 시각을 카운트하게 된다.

(3) 카운터 및 표시회로

카운터 및 표시회로는 그림 15 - 15와 같다. 여기서 IC 7490은 10진 카운터로 사용되며 Start/Stop 회로로부터 출력되는 펄스를 카운팅하고, 카운터의 출력은 다시 7 - Segment Driver IC 7447로 입력되어 카운터에서 출력되는 펄스를 디코딩하여 7 - Segment(FND)를 통해 0.1초부터 99.9초까지 표시해 주게 된다.

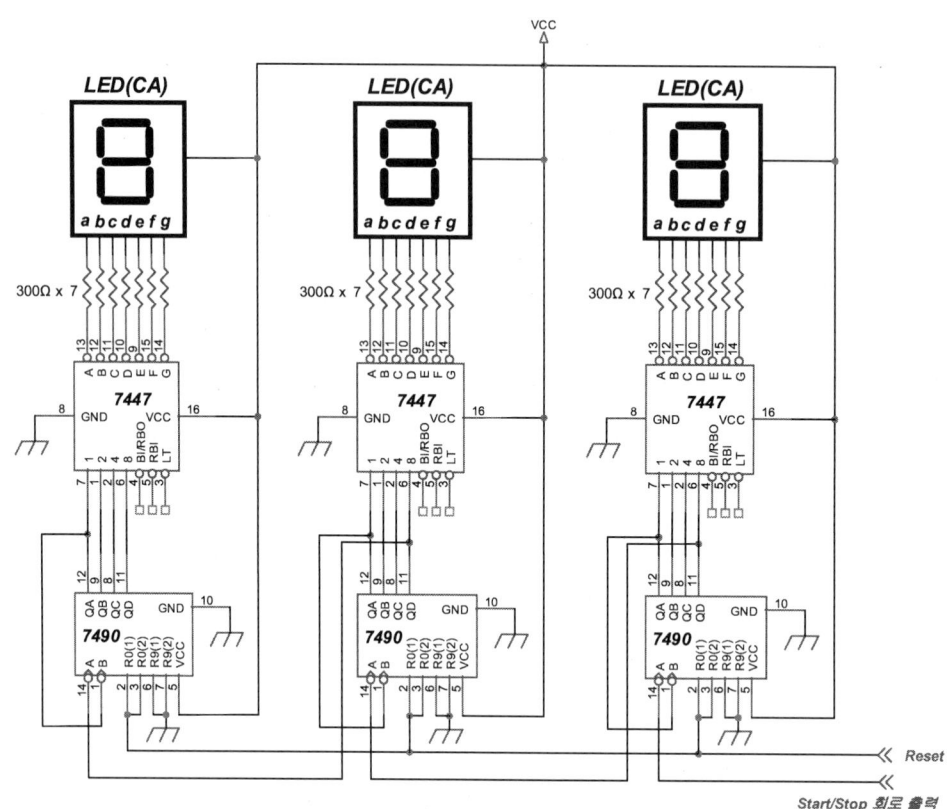

그림 15 - 15 카운터 및 표시회로

(4) Reset 회로

Reset 회로는 카운터를 Reset 시키는 기능을 수행하며, Reset 버튼과 7414 NOT 게이트를 통해 Start/Stop 회로의 카운터(IC 7490) 2, 3번 핀으로 High가 입력되어 표시기를 Reset 시키게 된다. 회로는 그림 15 – 16과 같다.

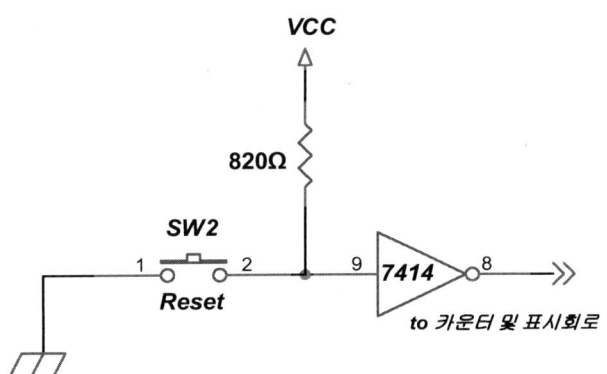

그림 15 – 16 Reset 회로

(1) 다음 그림 15－17의 회로를 구성하고 입력조건에 따른 출력 상태를 측정하여
     표 15－2에 기록하시오.

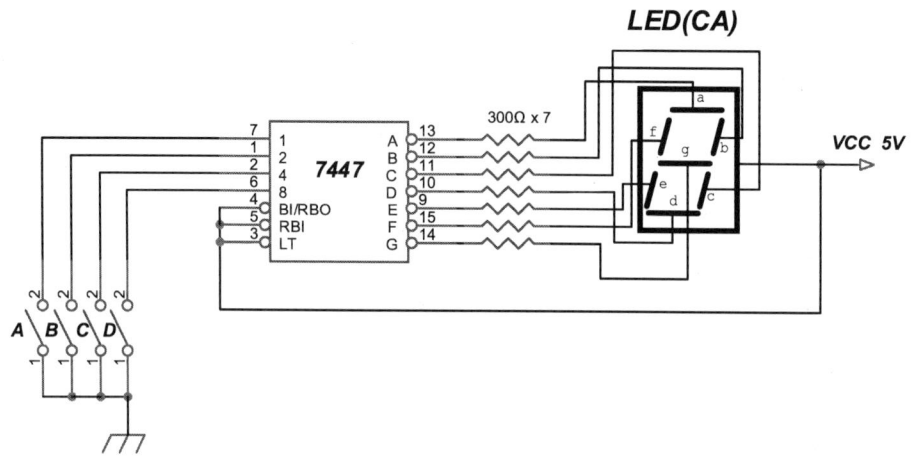

그림 15－17

표 15－2

D	C	B	A	표시
0	0	0	0	
0	0	0	1	
0	0	1	0	
0	0	1	1	
0	1	0	0	
0	1	0	1	
0	1	1	0	
0	1	1	1	
1	0	0	0	
1	0	0	1	

(2) 다음 그림 15 – 18의 회로를 구성하고 출력 상태를 측정하여 표 15 – 3에 기록하시오.

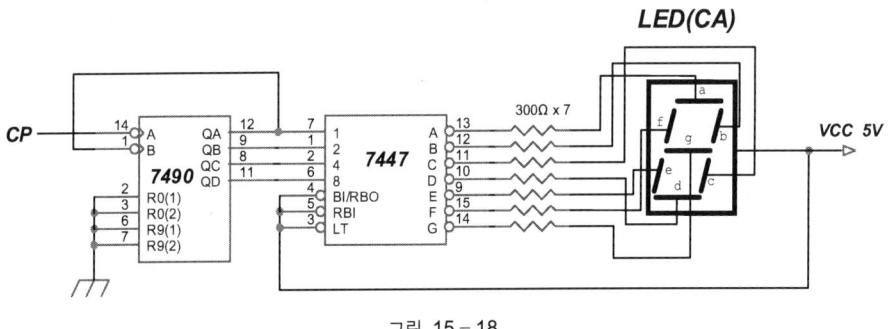

그림 15 – 18

표 15 – 3

표 15 – 4 74LS90 Reset/Count Table

Cp	표시
0	
1	
2	
3	
4	
5	
6	
7	
8	
9	

Reset Inputs				Output			
$R_{0(1)}$	$R_{0(2)}$	$R_{9(1)}$	$R_{9(2)}$	Qd	Qc	Qb	Qa
H	H	L	x	L	L	L	L
H	H	x	L	L	L	L	L
x	x	H	H	H	L	L	H
x	L	x	L	Count			
L	x	L	x	Count			
L	x	x	L	Count			
x	L	L	x	Count			

※BCD 카운트 시는 Qa와 입력 B를 연결

(3) 그림 15 – 19의 스톱워치 회로를 구성하고 기능이 정상으로 동작하는지 확인하시오.

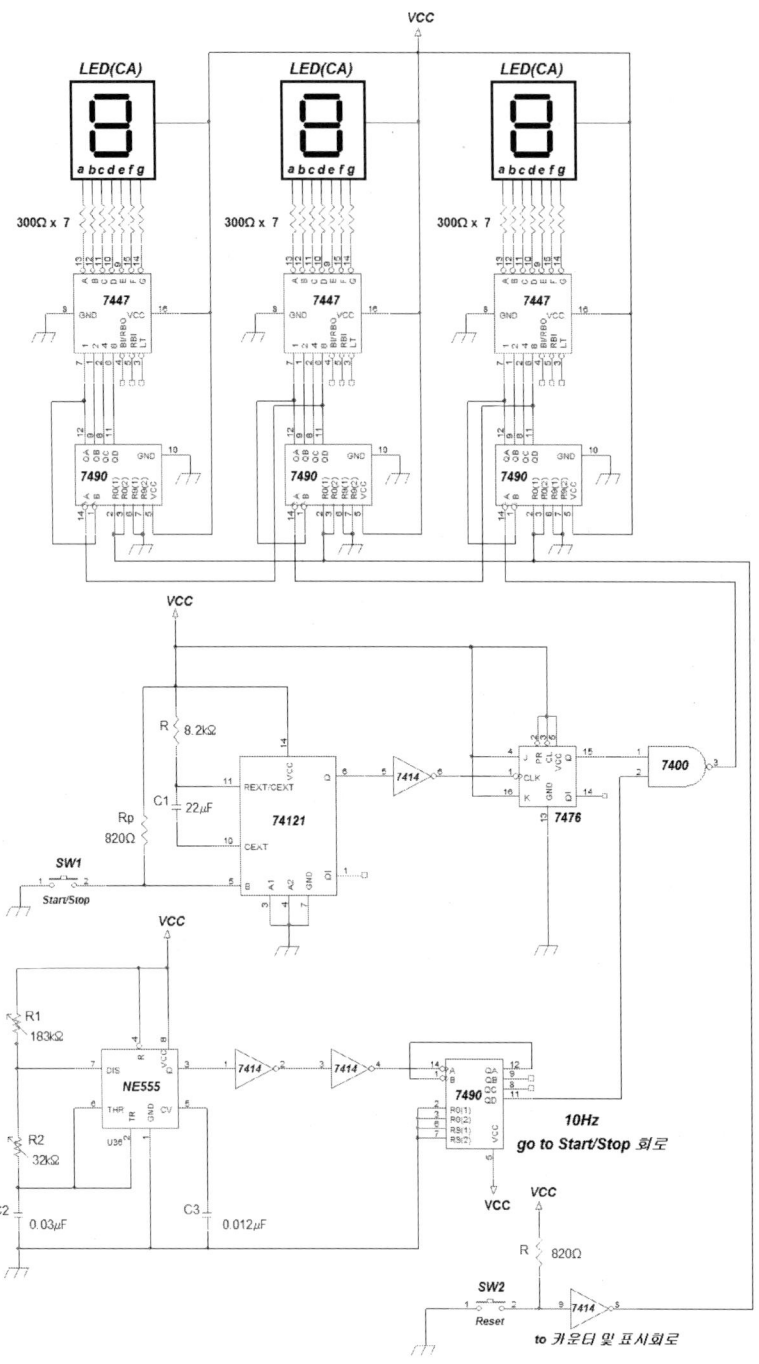

그림 15 – 19 스톱워치 회로

실험결과 Report	학과명	학번	성명
실험 15 LED와 카운터를 이용한 스톱워치			

☐ 실험 (1) ☐ 실험 (2)

표 15 - 2

D	C	B	A	표시
0	0	0	0	
0	0	0	1	
0	0	1	0	
0	0	1	1	
0	1	0	0	
0	1	0	1	
0	1	1	0	
0	1	1	1	
1	0	0	0	
1	0	0	1	

표 15 - 3

Cp	표시
0	
1	
2	
3	
4	
5	
6	
7	
8	
9	

☐ 실험 (3)

가. 스톱워치의 타이머 출력 주파수가 이론값과 일치하는지 확인하라.

나. Start/Stop 기능과 Reset 기능이 정상적인가?

찾아보기

부록 1. Electronic Component Datasheets

Timer   NE/SA/SE555/SE555C

DESCRIPTION

The 555 monolithic timing circuit is a highly stable controller capable of producing accurate time delays, or oscillation. In the time delay mode of operation, the time is precisely controlled by one external resistor and capacitor. For a stable operation as an oscillator, the free running frequency and the duty cycle are both accurately controlled with two external resistors and one capacitor. The circuit may be triggered and reset on falling waveforms, and the output structure can source or sink up to 200mA.

FEATURES

- Turn-off time less than 2μs
- Max. operating frequency greater than 500kHz
- Timing from microseconds to hours
- Operates in both astable and monostable modes
- High output current
- Adjustable duty cycle
- TTL compatible
- Temperature stability of 0.005% per °C

APPLICATIONS

- Precision timing
- Pulse generation
- Sequential timing
- Time delay generation
- Pulse width modulation

PIN CONFIGURATIONS

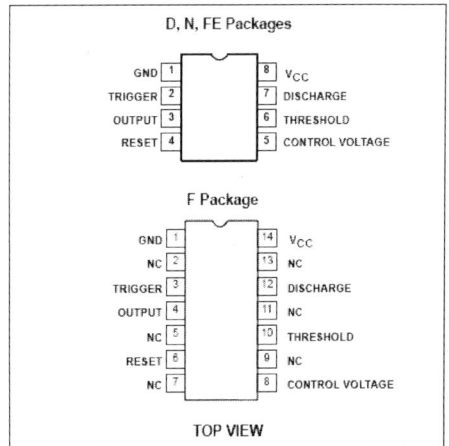

ORDERING INFORMATION

DESCRIPTION	TEMPERATURE RANGE	ORDER CODE	DWG #
8-Pin Plastic Small Outline (SO) Package	0 to +70°C	NE555D	0174C
8-Pin Plastic Dual In-Line Package (DIP)	0 to +70°C	NE555N	0404B
8-Pin Plastic Dual In-Line Package (DIP)	-40°C to +85°C	SA555N	0404B
8-Pin Plastic Small Outline (SO) Package	-40°C to +85°C	SA555D	0174C
8-Pin Hermetic Ceramic Dual In-Line Package (CERDIP)	-55°C to +125°C	SE555CFE	
8-Pin Plastic Dual In-Line Package (DIP)	-55°C to +125°C	SE555CN	0404B
14-Pin Plastic Dual In-Line Package (DIP)	-55°C to +125°C	SE555N	0405B
8-Pin Hermetic Cerdip	-55°C to +125°C	SE555FE	
14-Pin Ceramic Dual In-Line Package (CERDIP)	0 to +70°C	NE555F	0581B
14-Pin Ceramic Dual In-Line Package (CERDIP)	-55°C to +125°C	SE555F	0581B
14-Pin Ceramic Dual In-Line Package (CERDIP)	-55°C to +125°C	SE555CF	0581B

BLOCK DIAGRAM

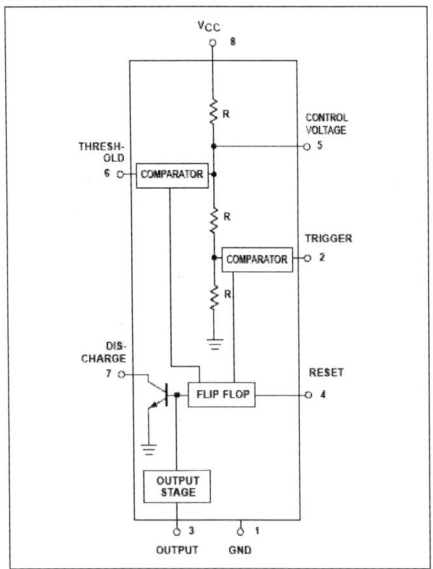

EQUIVALENT SCHEMATIC

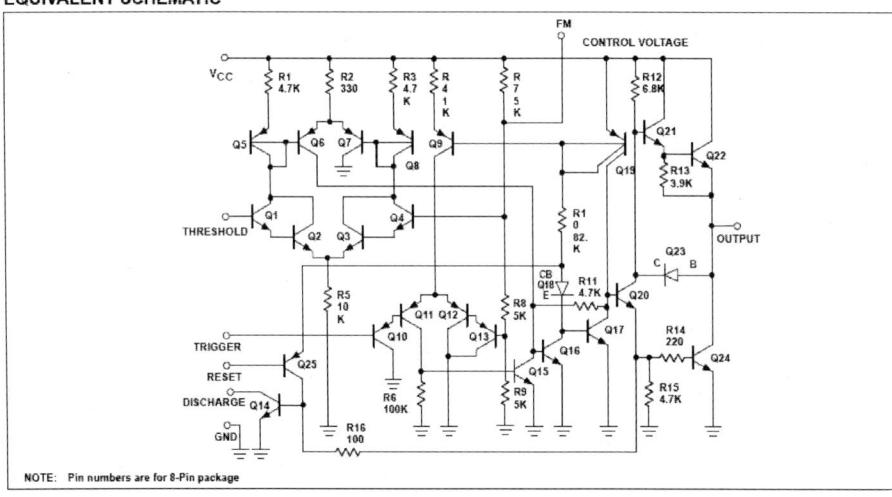

NOTE: Pin numbers are for 8-Pin package

ABSOLUTE MAXIMUM RATINGS

SYMBOL	PARAMETER	RATING	UNIT
V_{CC}	Supply voltage		
	SE555	+18	V
	NE555, SE555C, SA555	+16	V
P_D	Maximum allowable power dissipation[1]	600	mW
T_A	Operating ambient temperature range		
	NE555	0 to +70	°C
	SA555	-40 to +85	°C
	SE555, SE555C	-55 to +125	°C
T_{STG}	Storage temperature range	-65 to +150	°C
T_{SOLD}	Lead soldering temperature (10sec max)	+300	°C

NOTES:
1. The junction temperature must be kept below 125°C for the D package and below 150°C for the FE, N and F packages. At ambient tempera-
tures above 25°C, where this limit would be derated by the following factors.
   D package 160°C/W
   FE package 150°C/W
   N package 100°C/W
   F package 105°C/W

DC AND AC ELECTRICAL CHARACTERISTICS

T_A = 25°C, V_{CC} = +5V to +15 unless otherwise specified.

SYMBOL	PARAMETER	TEST CONDITIONS	SE555			NE555/SE555C			UNIT
			Min	Typ	Max	Min	Typ	Max	
V_{CC}	Supply voltage		4.5		18	4.5		16	V
I_{CC}	Supply current (low	V_{CC}=5V, R_L=∞		3	5		3	6	mA
	state)[1]	V_{CC}=15V, R_L=∞		10	12		10	15	mA
	Timing error (monostable)	R_A=2kΩ to 100kΩ							
t_M	Initial accuracy[2]	C=0.1µF		0.5	2.0		1.0	3.0	%
$\Delta t_M/\Delta T$	Drift with temperature			30	100		50	150	ppm/°C
$\Delta t_M/\Delta V_S$	Drift with supply voltage			0.05	0.2		0.1	0.5	%/V
	Timing error (astable)	R_A, R_B=1kΩ to 100kΩ							
t_A	Initial accuracy[2]	C=0.1µF		4	6		5	13	%
$\Delta t_A/\Delta T$	Drift with temperature	V_{CC}=15V			500			500	ppm/°C
$\Delta t_A/\Delta V_S$	Drift with supply voltage			0.15	0.6		0.3	1	%/V
V_C	Control voltage level	V_{CC}=15V	9.6	10.0	10.4	9.0	10.0	11.0	V
		V_{CC}=5V	2.9	3.33	3.8	2.6	3.33	4.0	V
V_{TH}	Threshold voltage	V_{CC}=15V	9.4	10.0	10.6	8.8	10.0	11.2	V
		V_{CC}=5V	2.7	3.33	4.0	2.4	3.33	4.2	V
I_{TH}	Threshold current[3]			0.1	0.25		0.1	0.25	µA
V_{TRIG}	Trigger voltage	V_{CC}=15V	4.8	5.0	5.2	4.5	5.0	5.6	V
		V_{CC}=5V	1.45	1.67	1.9	1.1	1.67	2.2	V
I_{TRIG}	Trigger current	V_{TRIG}=0V		0.5	0.9		0.5	2.0	µA
V_{RESET}	Reset voltage[4]	V_{CC}=15V, V_{TH}=10.5V	0.3		1.0	0.3		1.0	V
I_{RESET}	Reset current	V_{RESET}=0.4V		0.1	0.4		0.1	0.4	mA
	Reset current	V_{RESET}=0V		0.4	1.0		0.4	1.5	mA
V_{OL}	Output voltage (low)	V_{CC}=15V							
		I_{SINK}=10mA		0.1	0.15		0.1	0.25	V
		I_{SINK}=50mA		0.4	0.5		0.4	0.75	V
		I_{SINK}=100mA		2.0	2.2		2.0	2.5	V
		I_{SINK}=200mA		2.5			2.5		V
		V_{CC}=5V							
		I_{SINK}=8mA		0.1	0.25		0.3	0.4	V
		I_{SINK}=5mA		0.05	0.2		0.25	0.35	V
V_{OH}	Output voltage (high)	V_{CC}=15V							
		I_{SOURCE}=200mA		12.5			12.5		V
		I_{SOURCE}=100mA	13.0	13.3		12.75	13.3		V
		V_{CC}=5V							
		I_{SOURCE}=100mA	3.0	3.3		2.75	3.3		V
t_{OFF}	Turn-off time[5]	V_{RESET}=V_{CC}		0.5	2.0		0.5	2.0	µs
t_R	Rise time of output			100	200		100	300	ns
t_F	Fall time of output			100	200		100	300	ns
	Discharge leakage current			20	100		20	100	nA

NOTES:
1. Supply current when output high typically 1mA less.
2. Tested at V_{CC}=5V and V_{CC}=15V.
3. This will determine the max value of R_A+R_B, for 15V operation, the max total R=10MΩ, and for 5V operation, the max. total R=3.4MΩ.
4. Specified with trigger input high.
5. Time measured from a positive going input pulse from 0 to 0.8×V_{CC} into the threshold to the drop from high to low of the output. Trigger is tied to threshold.

TYPICAL APPLICATIONS

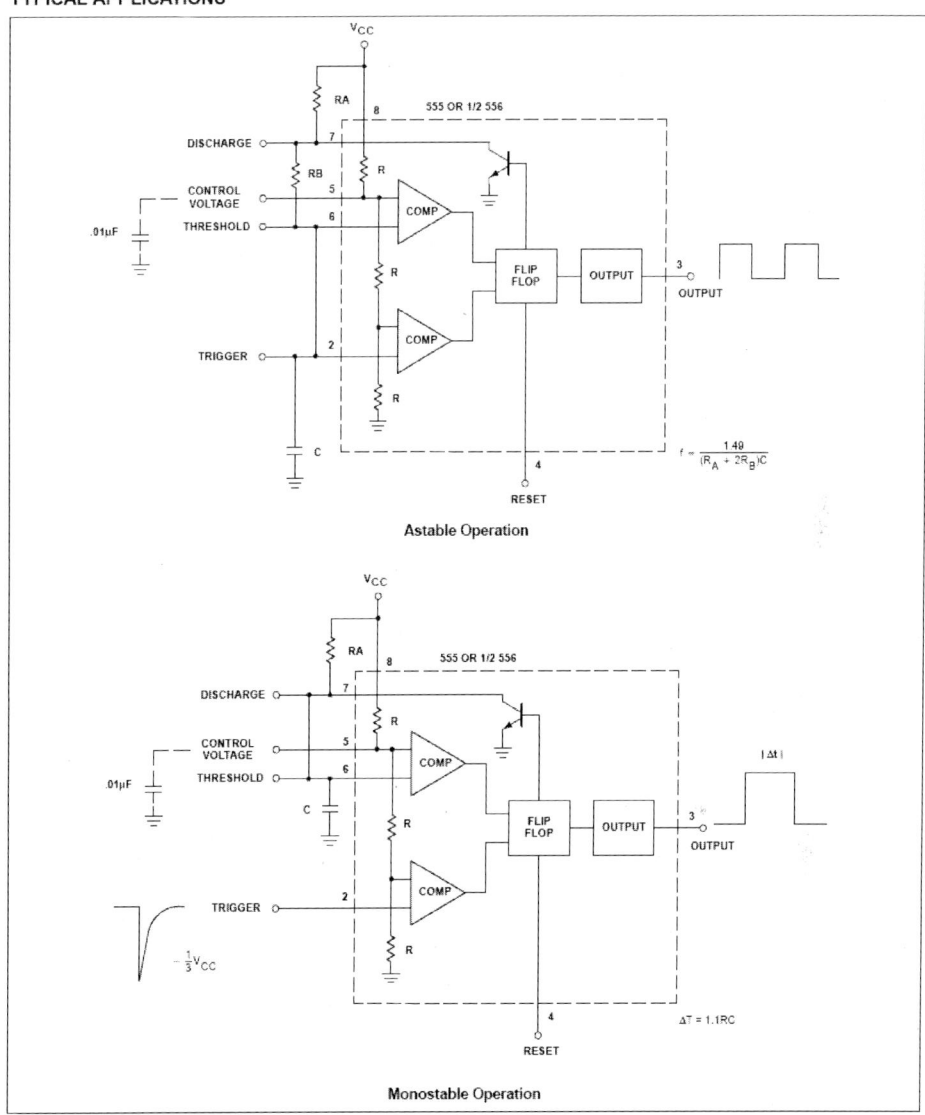

Astable Operation

Monostable Operation

SN74LS00

Quad 2-Input NAND Gate

- ESD > 3500 Volts

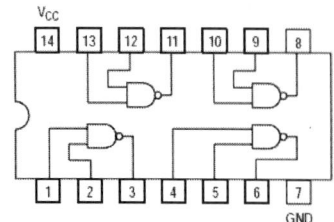

ON Semiconductor
Formerly a Division of Motorola
http://onsemi.com

LOW
POWER
SCHOTTKY

GUARANTEED OPERATING RANGES

Symbol	Parameter	Min	Typ	Max	Unit
V_{CC}	Supply Voltage	4.75	5.0	5.25	V
T_A	Operating Ambient Temperature Range	0	25	70	°C
I_{OH}	Output Current – High			–0.4	mA
I_{OL}	Output Current – Low			8.0	mA

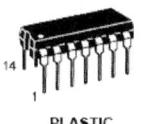

PLASTIC
N SUFFIX
CASE 646

SOIC
D SUFFIX
CASE 751A

ORDERING INFORMATION

Device	Package	Shipping
SN74LS00N	14 Pin DIP	2000 Units/Box
SN74LS00D	14 Pin	2500/Tape & Reel

DC CHARACTERISTICS OVER OPERATING TEMPERATURE RANGE (unless otherwise specified)

Symbol	Parameter	Limits			Unit	Test Conditions	
		Min	Typ	Max			
V_{IH}	Input HIGH Voltage	2.0			V	Guaranteed Input HIGH Voltage for All Inputs	
V_{IL}	Input LOW Voltage			0.8	V	Guaranteed Input LOW Voltage for All Inputs	
V_{IK}	Input Clamp Diode Voltage		−0.65	−1.5	V	V_{CC} = MIN, I_{IN} = −18 mA	
V_{OH}	Output HIGH Voltage	2.7	3.5		V	V_{CC} = MIN, I_{OH} = MAX, V_{IN} = V_{IH} or V_{IL} per Truth Table	
V_{OL}	Output LOW Voltage		0.25	0.4	V	I_{OL} = 4.0 mA	V_{CC} = V_{CC} MIN, V_{IN} = V_{IL} or V_{IH} per Truth Table
			0.35	0.5	V	I_{OL} = 8.0 mA	
I_{IH}	Input HIGH Current			20	μA	V_{CC} = MAX, V_{IN} = 2.7 V	
				0.1	mA	V_{CC} = MAX, V_{IN} = 7.0 V	
I_{IL}	Input LOW Current			−0.4	mA	V_{CC} = MAX, V_{IN} = 0.4 V	
I_{OS}	Short Circuit Current (Note 1)	−20		−100	mA	V_{CC} = MAX	
I_{CC}	Power Supply Current Total, Output HIGH			1.6	mA	V_{CC} = MAX	
	Total, Output LOW			4.4			

Note 1: Not more than one output should be shorted at a time, nor for more than 1 second.

AC CHARACTERISTICS (T_A = 25°C)

Symbol	Parameter	Limits			Unit	Test Conditions
		Min	Typ	Max		
t_{PLH}	Turn–Off Delay, Input to Output		9.0	15	ns	V_{CC} = 5.0 V
t_{PHL}	Turn–On Delay, Input to Output		10	15	ns	C_L = 15 pF

 **MOTOROLA**

QUAD 2-INPUT NOR GATE

SN54/74LS02

QUAD 2-INPUT NOR GATE

LOW POWER SCHOTTKY

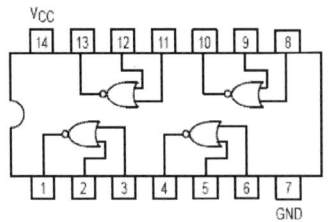

Vcc
14 13 12 11 10 9 8

1 2 3 4 5 6 7
GND

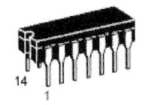

J SUFFIX
CERAMIC
CASE 632-08

14 1

N SUFFIX
PLASTIC
CASE 646-06

14 1

D SUFFIX
SOIC
CASE 751A-02

14 1

ORDERING INFORMATION

SN54LSXXJ   Ceramic
SN74LSXXN   Plastic
SN74LSXXD   SOIC

GUARANTEED OPERATING RANGES

Symbol	Parameter			Min	Typ	Max	Unit
V_{CC}	Supply Voltage		54	4.5	5.0	5.5	V
			74	4.75	5.0	5.25	
T_A	Operating Ambient Temperature Range		54	−55	25	125	°C
			74	0	25	70	
I_{OH}	Output Current — High		54, 74			−0.4	mA
I_{OL}	Output Current — Low		54			4.0	mA
			74			8.0	

DC CHARACTERISTICS OVER OPERATING TEMPERATURE RANGE (unless otherwise specified)

Symbol	Parameter		Min	Typ	Max	Unit	Test Conditions	
V_{IH}	Input HIGH Voltage		2.0			V	Guaranteed Input HIGH Voltage for All Inputs	
V_{IL}	Input LOW Voltage	54			0.7	V	Guaranteed Input LOW Voltage for All Inputs	
		74			0.8			
V_{IK}	Input Clamp Diode Voltage			−0.65	−1.5	V	V_{CC} = MIN, I_{IN} = − 18 mA	
V_{OH}	Output HIGH Voltage	54	2.5	3.5		V	V_{CC} = MIN, I_{OH} = MAX, V_{IN} = V_{IH} or V_{IL} per Truth Table	
		74	2.7	3.5		V		
V_{OL}	Output LOW Voltage	54, 74		0.25	0.4	V	I_{OL} = 4.0 mA	V_{CC} = V_{CC} MIN, V_{IN} = V_{IL} or V_{IH} per Truth Table
		74		0.35	0.5	V	I_{OL} = 8.0 mA	
I_{IH}	Input HIGH Current				20	µA	V_{CC} = MAX, V_{IN} = 2.7 V	
					0.1	mA	V_{CC} = MAX, V_{IN} = 7.0 V	
I_{IL}	Input LOW Current				−0.4	mA	V_{CC} = MAX, V_{IN} = 0.4 V	
I_{OS}	Short Circuit Current (Note 1)		−20		−100	mA	V_{CC} = MAX	
I_{CC}	Power Supply Current Total, Output HIGH				3.2	mA	V_{CC} = MAX	
	Total, Output LOW				5.4			

Note 1: Not more than one output should be shorted at a time, nor for more than 1 second.

AC CHARACTERISTICS (T_A = 25°C)

Symbol	Parameter	Min	Typ	Max	Unit	Test Conditions
t_{PLH}	Turn-Off Delay, Input to Output		10	15	ns	V_{CC} = 5.0 V C_L = 15 pF
t_{PHL}	Turn-On Delay, Input to Output		10	15	ns	

SN74LS04

Hex Inverter

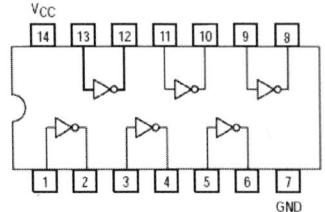

GND

ON Semiconductor

Formerly a Division of Motorola

http://onsemi.com

LOW
POWER
SCHOTTKY

GUARANTEED OPERATING RANGES

Symbol	Parameter	Min	Typ	Max	Unit
V_{CC}	Supply Voltage	4.75	5.0	5.25	V
T_A	Operating Ambient Temperature Range	0	25	70	°C
I_{OH}	Output Current – High			−0.4	mA
I_{OL}	Output Current – Low			8.0	mA

PLASTIC
N SUFFIX
CASE 646

SOIC
D SUFFIX
CASE 751A

ORDERING INFORMATION

Device	Package	Shipping
SN74LS04N	14 Pin DIP	2000 Units/Box
SN74LS04D	14 Pin	2500/Tape & Reel

DC CHARACTERISTICS OVER OPERATING TEMPERATURE RANGE (unless otherwise specified)

Symbol	Parameter	Limits			Unit	Test Conditions	
		Min	Typ	Max			
V_{IH}	Input HIGH Voltage	2.0			V	Guaranteed Input HIGH Voltage for All Inputs	
V_{IL}	Input LOW Voltage			0.8	V	Guaranteed Input LOW Voltage for All Inputs	
V_{IK}	Input Clamp Diode Voltage		−0.65	−1.5	V	V_{CC} = MIN, I_{IN} = −18 mA	
V_{OH}	Output HIGH Voltage	2.7	3.5		V	V_{CC} = MIN, I_{OH} = MAX, V_{IN} = V_{IH} or V_{IL} per Truth Table	
V_{OL}	Output LOW Voltage		0.25	0.4	V	I_{OL} = 4.0 mA	V_{CC} = V_{CC} MIN, V_{IN} = V_{IL} or V_{IH} per Truth Table
			0.35	0.5	V	I_{OL} = 8.0 mA	
I_{IH}	Input HIGH Current			20	μA	V_{CC} = MAX, V_{IN} = 2.7 V	
				0.1	mA	V_{CC} = MAX, V_{IN} = 7.0 V	
I_{IL}	Input LOW Current			−0.4	mA	V_{CC} = MAX, V_{IN} = 0.4 V	
I_{OS}	Short Circuit Current (Note 1)	−20		−100	mA	V_{CC} = MAX	
I_{CC}	Power Supply Current Total, Output HIGH			2.4	mA	V_{CC} = MAX	
	Total, Output LOW			6.6			

Note 1: Not more than one output should be shorted at a time, nor for more than 1 second.

AC CHARACTERISTICS (T_A = 25°C)

Symbol	Parameter	Limits			Unit	Test Conditions
		Min	Typ	Max		
t_{PLH}	Turn–Off Delay, Input to Output		9.0	15	ns	V_{CC} = 5.0 V
t_{PHL}	Turn–On Delay, Input to Output		10	15	ns	C_L = 15 pF

SN74LS05

Hex Inverter

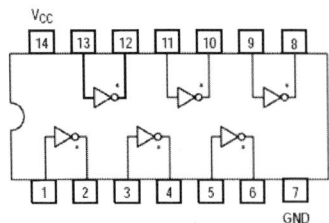

*OPEN COLLECTOR OUTPUTS

ON Semiconductor
Formerly a Division of Motorola
http://onsemi.com

LOW
POWER
SCHOTTKY

PLASTIC
N SUFFIX
CASE 646

SOIC
D SUFFIX
CASE 751A

GUARANTEED OPERATING RANGES

Symbol	Parameter	Min	Typ	Max	Unit
V_{CC}	Supply Voltage	4.75	5.0	5.25	V
T_A	Operating Ambient Temperature Range	0	25	70	°C
V_{OH}	Output Voltage – High			5.5	V
I_{OL}	Output Current – Low			8.0	mA

ORDERING INFORMATION

Device	Package	Shipping
SN74LS05N	14 Pin DIP	2000 Units/Box
SN74LS05D	14 Pin	2500/Tape & Reel

DC CHARACTERISTICS OVER OPERATING TEMPERATURE RANGE (unless otherwise specified)

Symbol	Parameter	Limits			Unit	Test Conditions	
		Min	Typ	Max			
V_{IH}	Input HIGH Voltage	2.0			V	Guaranteed Input HIGH Voltage for All Inputs	
V_{IL}	Input LOW Voltage			0.8	V	Guaranteed Input LOW Voltage for All Inputs	
V_{IK}	Input Clamp Diode Voltage		−0.65	−1.5	V	V_{CC} = MIN, I_{IN} = −18 mA	
I_{OH}	Output HIGH Current			100	µA	V_{CC} = MIN, V_{OH} = MAX	
V_{OL}	Output LOW Voltage		0.25	0.4	V	I_{OL} = 4.0 mA	V_{CC} = V_{CC} MIN, V_{IN} = V_{IL} or V_{IH} per Truth Table
			0.35	0.5	V	I_{OL} = 8.0 mA	
I_{IH}	Input HIGH Current			20	µA	V_{CC} = MAX, V_{IN} = 2.7 V	
				0.1	mA	V_{CC} = MAX, V_{IN} = 7.0 V	
I_{IL}	Input LOW Current			−0.4	mA	V_{CC} = MAX, V_{IN} = 0.4 V	
I_{CC}	Power Supply Current Total, Output HIGH			2.4	mA	V_{CC} = MAX	
	Total, Output LOW			6.6			

AC CHARACTERISTICS (T_A = 25°C)

Symbol	Parameter	Limits			Unit	Test Conditions
		Min	Typ	Max		
t_{PLH}	Turn–Off Delay, Input to Output		17	32	ns	V_{CC} = 5.0 V C_L = 15 pF, R_L = 2.0 kΩ
t_{PHL}	Turn–On Delay, Input to Output		15	28	ns	

SN74LS08

Quad 2-Input AND Gate

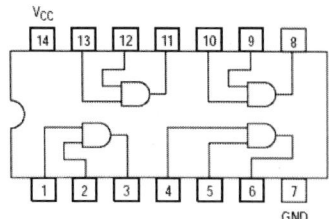

ON Semiconductor
Formerly a Division of Motorola
http://onsemi.com

LOW
POWER
SCHOTTKY

GUARANTEED OPERATING RANGES

Symbol	Parameter	Min	Typ	Max	Unit
V_{CC}	Supply Voltage	4.75	5.0	5.25	V
T_A	Operating Ambient Temperature Range	0	25	70	°C
I_{OH}	Output Current – High			−0.4	mA
I_{OL}	Output Current – Low			8.0	mA

PLASTIC
N SUFFIX
CASE 646

SOIC
D SUFFIX
CASE 751A

ORDERING INFORMATION

Device	Package	Shipping
SN74LS08N	14 Pin DIP	2000 Units/Box
SN74LS08D	14 Pin	2500/Tape & Reel

DC CHARACTERISTICS OVER OPERATING TEMPERATURE RANGE (unless otherwise specified)

Symbol	Parameter	Limits			Unit	Test Conditions	
		Min	Typ	Max			
V_{IH}	Input HIGH Voltage	2.0			V	Guaranteed Input HIGH Voltage for All Inputs	
V_{IL}	Input LOW Voltage			0.8	V	Guaranteed Input LOW Voltage for All Inputs	
V_{IK}	Input Clamp Diode Voltage		−0.65	−1.5	V	V_{CC} = MIN, I_{IN} = −18 mA	
V_{OH}	Output HIGH Voltage	2.7	3.5		V	V_{CC} = MIN, I_{OH} = MAX, V_{IN} = V_{IH} or V_{IL} per Truth Table	
V_{OL}	Output LOW Voltage		0.25	0.4	V	I_{OL} = 4.0 mA	V_{CC} = V_{CC} MIN, V_{IN} = V_{IL} or V_{IH} per Truth Table
			0.35	0.5	V	I_{OL} = 8.0 mA	
I_{IH}	Input HIGH Current			20	μA	V_{CC} = MAX, V_{IN} = 2.7 V	
				0.1	mA	V_{CC} = MAX, V_{IN} = 7.0 V	
I_{IL}	Input LOW Current			−0.4	mA	V_{CC} = MAX, V_{IN} = 0.4 V	
I_{OS}	Short Circuit Current (Note 1)	−20		−100	mA	V_{CC} = MAX	
I_{CC}	Power Supply Current Total, Output HIGH			4.8	mA	V_{CC} = MAX	
	Total, Output LOW			8.8			

Note 1: Not more than one output should be shorted at a time, nor for more than 1 second.

AC CHARACTERISTICS (T_A = 25°C)

Symbol	Parameter	Limits			Unit	Test Conditions
		Min	Typ	Max		
t_{PLH}	Turn–Off Delay, Input to Output		8.0	15	ns	V_{CC} = 5.0 V C_L = 15 pF
t_{PHL}	Turn–On Delay, Input to Output		10	20	ns	

 **MOTOROLA**

TRIPLE 3-INPUT NAND GATE

SN54/74LS10

TRIPLE 3-INPUT NAND GATE
LOW POWER SCHOTTKY

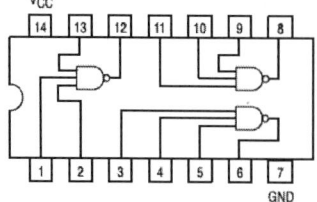

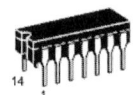

J SUFFIX
CERAMIC
CASE 632-08

N SUFFIX
PLASTIC
CASE 646-06

D SUFFIX
SOIC
CASE 751A-02

ORDERING INFORMATION

SN54LSXXJ   Ceramic
SN74LSXXN   Plastic
SN74LSXXD   SOIC

GUARANTEED OPERATING RANGES

Symbol	Parameter		Min	Typ	Max	Unit
V$_{CC}$	Supply Voltage	54	4.5	5.0	5.5	V
		74	4.75	5.0	5.25	
T$_A$	Operating Ambient Temperature Range	54	−55	25	125	°C
		74	0	25	70	
I$_{OH}$	Output Current — High	54, 74			−0.4	mA
I$_{OL}$	Output Current — Low	54			4.0	mA
		74			8.0	

DC CHARACTERISTICS OVER OPERATING TEMPERATURE RANGE (unless otherwise specified)

Symbol	Parameter		Min	Typ	Max	Unit	Test Conditions	
				Limits				
V_{IH}	Input HIGH Voltage		2.0			V	Guaranteed Input HIGH Voltage for All Inputs	
V_{IL}	Input LOW Voltage	54			0.7	V	Guaranteed Input LOW Voltage for All Inputs	
		74			0.8			
V_{IK}	Input Clamp Diode Voltage			−0.65	−1.5	V	V_{CC} = MIN, I_{IN} = −18 mA	
V_{OH}	Output HIGH Voltage	54	2.5	3.5		V	V_{CC} = MIN, I_{OH} = MAX, V_{IN} = V_{IH} or V_{IL} per Truth Table	
		74	2.7	3.5		V		
V_{OL}	Output LOW Voltage	54, 74		0.25	0.4	V	I_{OL} = 4.0 mA	V_{CC} = V_{CC} MIN, V_{IN} = V_{IL} or V_{IH} per Truth Table
		74		0.35	0.5	V	I_{OL} = 8.0 mA	
I_{IH}	Input HIGH Current				20	μA	V_{CC} = MAX, V_{IN} = 2.7 V	
					0.1	mA	V_{CC} = MAX, V_{IN} = 7.0 V	
I_{IL}	Input LOW Current				−0.4	mA	V_{CC} = MAX, V_{IN} = 0.4 V	
I_{OS}	Short Circuit Current (Note 1)		−20		−100	mA	V_{CC} = MAX	
I_{CC}	Power Supply Current Total, Output HIGH				1.2	mA	V_{CC} = MAX	
	Total, Output LOW				3.3			

Note 1: Not more than one output should be shorted at a time, nor for more than 1 second.

AC CHARACTERISTICS (T_A = 25°C)

Symbol	Parameter	Min	Typ	Max	Unit	Test Conditions
			Limits			
t_{PLH}	Turn-Off Delay, Input to Output		9.0	15	ns	V_{CC} = 5.0 V C_L = 15 pF
t_{PHL}	Turn-On Delay, Input to Output		10	15	ns	

 **MOTOROLA**

TRIPLE 3-INPUT AND GATE

TRIPLE 3-INPUT AND GATE

LOW POWER SCHOTTKY

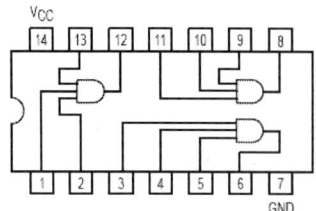

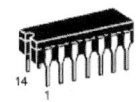

J SUFFIX
CERAMIC
CASE 632-08

N SUFFIX
PLASTIC
CASE 646-06

D SUFFIX
SOIC
CASE 751A-02

ORDERING INFORMATION

SN54LSXXJ   Ceramic
SN74LSXXN   Plastic
SN74LSXXD   SOIC

GUARANTEED OPERATING RANGES

Symbol	Parameter		Min	Typ	Max	Unit
V_{CC}	Supply Voltage	54	4.5	5.0	5.5	V
		74	4.75	5.0	5.25	
T_A	Operating Ambient Temperature Range	54	−55	25	125	°C
		74	0	25	70	
I_{OH}	Output Current — High	54, 74			−0.4	mA
I_{OL}	Output Current — Low	54			4.0	mA
		74			8.0	

DC CHARACTERISTICS OVER OPERATING TEMPERATURE RANGE (unless otherwise specified)

Symbol	Parameter		Min	Typ	Max	Unit	Test Conditions	
V_{IH}	Input HIGH Voltage		2.0			V	Guaranteed Input HIGH Voltage for All Inputs	
V_{IL}	Input LOW Voltage	54			0.7	V	Guaranteed Input LOW Voltage for All Inputs	
		74			0.8			
V_{IK}	Input Clamp Diode Voltage			−0.65	−1.5	V	V_{CC} = MIN, I_{IN} = −18 mA	
V_{OH}	Output HIGH Voltage	54	2.5	3.5		V	V_{CC} = MIN, I_{OH} = MAX, V_{IN} = V_{IH} or V_{IL} per Truth Table	
		74	2.7	3.5		V		
V_{OL}	Output LOW Voltage	54, 74		0.25	0.4	V	I_{OL} = 4.0 mA	V_{CC} = V_{CC} MIN, V_{IN} = V_{IL} or V_{IH} per Truth Table
		74		0.35	0.5	V	I_{OL} = 8.0 mA	
I_{IH}	Input HIGH Current				20	μA	V_{CC} = MAX, V_{IN} = 2.7 V	
					0.1	mA	V_{CC} = MAX, V_{IN} = 7.0 V	
I_{IL}	Input LOW Current				−0.4	mA	V_{CC} = MAX, V_{IN} = 0.4 V	
I_{OS}	Short Circuit Current (Note 1)		−20		−100	mA	V_{CC} = MAX	
I_{CC}	Power Supply Current Total, Output HIGH				3.6	mA	V_{CC} = MAX	
	Total, Output LOW				6.6			

Note 1: Not more than one output should be shorted at a time, nor for more than 1 second.

AC CHARACTERISTICS (T_A = 25°C)

Symbol	Parameter	Min	Typ	Max	Unit	Test Conditions
t_{PLH}	Turn-Off Delay, Input to Output		8.0	15	ns	V_{CC} = 5.0 V C_L = 15 pF
t_{PHL}	Turn-On Delay, Input to Output		10	20	ns	

SN74LS14

Schmitt Triggers
Dual Gate/Hex Inverter

The SN74LS14 contains logic gates/inverters which accept standard TTL input signals and provide standard TTL output levels. They are capable of transforming slowly changing input signals into sharply defined, jitter-free output signals. Additionally, they have greater noise margin than conventional inverters.

Each circuit contains a Schmitt trigger followed by a Darlington level shifter and a phase splitter driving a TTL totem pole output. The Schmitt trigger uses positive feedback to effectively speed-up slow input transitions, and provide different input threshold voltages for positive and negative-going transitions. This hysteresis between the positive-going and negative-going input thresholds (typically 800 mV) is determined internally by resistor ratios and is essentially insensitive to temperature and supply voltage variations.

ON Semiconductor™

http://onsemi.com

**LOW
POWER
SCHOTTKY**

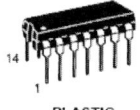

PLASTIC
N SUFFIX
CASE 646

SOIC
D SUFFIX
CASE 751A

LOGIC AND CONNECTION DIAGRAMS

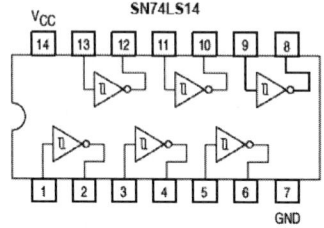

GUARANTEED OPERATING RANGES

Symbol	Parameter	Min	Typ	Max	Unit
V_{CC}	Supply Voltage	4.75	5.0	5.25	V
T_A	Operating Ambient Temperature Range	0	25	70	°C
I_{OH}	Output Current – High			−0.4	mA
I_{OL}	Output Current – Low			8.0	mA

ORDERING INFORMATION

Device	Package	Shipping
SN74LS14N	14 Pin DIP	2000 Units/Box
SN74LS14D	SOIC–14	55 Units/Rail
SN74LS14DR2	SOIC–14	2500/Tape & Reel

DC CHARACTERISTICS OVER OPERATING TEMPERATURE RANGE (unless otherwise specified)

Symbol	Parameter	Limits			Unit	Test Conditions
		Min	Typ	Max		
V_{T+}	Positive-Going Threshold Voltage	1.5		2.0	V	$V_{CC} = 5.0$ V
V_{T-}	Negative-Going Threshold Voltage	0.6		1.1	V	$V_{CC} = 5.0$ V
$V_{T+}-V_{T-}$	Hysteresis	0.4	0.8		V	$V_{CC} = 5.0$ V
V_{IK}	Input Clamp Diode Voltage		−0.65	−1.5	V	$V_{CC} = $ MIN, $I_{IN} = -18$ mA
V_{OH}	Output HIGH Voltage	2.7	3.4		V	$V_{CC} = $ MIN, $I_{OH} = -400$ µA, $V_{IN} = V_{IL}$
V_{OL}	Output LOW Voltage		0.25	0.4	V	$V_{CC} = $ MIN, $I_{OL} = 4.0$ mA, $V_{IN} = 2.0$ V
			0.35	0.5	V	$V_{CC} = $ MIN, $I_{OL} = 8.0$ mA, $V_{IN} = 2.0$ V
I_{T+}	Input Current at Positive-Going Threshold		−0.14		mA	$V_{CC} = 5.0$ V, $V_{IN} = V_{T+}$
I_{T-}	Input Current at Negative-Going Threshold		−0.18		mA	$V_{CC} = 5.0$ V, $V_{IN} = V_{T-}$
I_{IH}	Input HIGH Current		1.0	20	µA	$V_{CC} = $ MAX, $V_{IN} = 2.7$ V
				0.1	mA	$V_{CC} = $ MAX, $V_{IN} = 7.0$ V
I_{IL}	Input LOW Current			−0.4	mA	$V_{CC} = $ MAX, $V_{IN} = 0.4$ V
I_{OS}	Short Circuit Current (Note 1)	−20		−100	mA	$V_{CC} = $ MAX, $V_{OUT} = 0$ V
I_{CC}	Power Supply Current Total, Output HIGH Total, Output LOW		8.6	16	mA	$V_{CC} = $ MAX
			12	21		

Note 1: Not more than one output should be shorted at a time, nor for more than 1 second.

AC CHARACTERISTICS ($T_A = 25°C$)

Symbol	Parameter	Max	Unit	Test Conditions
t_{PLH}	Propagation Delay, Input to Output	22	ns	$V_{CC} = 5.0$ V
t_{PHL}	Propagation Delay, Input to Output	22	ns	$C_L = 15$ pF

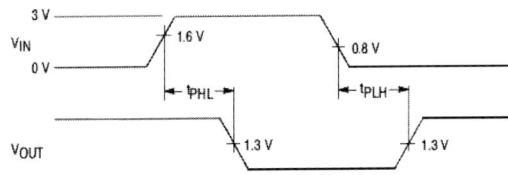

Figure 1. AC Waveforms

TRIPLE 3-INPUT AND GATE

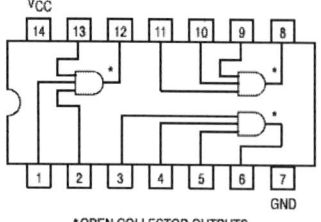

Vcc

14 13 12 11 10 9 8

*

*

*

1 2 3 4 5 6 7
GND

*OPEN COLLECTOR OUTPUTS

SN54/74LS15

TRIPLE 3-INPUT AND GATE

LOW POWER SCHOTTKY

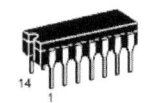

J SUFFIX
CERAMIC
CASE 632-08

N SUFFIX
PLASTIC
CASE 646-06

D SUFFIX
SOIC
CASE 751A-02

ORDERING INFORMATION

SN54LSXXJ   Ceramic
SN74LSXXN   Plastic
SN74LSXXD   SOIC

GUARANTEED OPERATING RANGES

Symbol	Parameter		Min	Typ	Max	Unit
V_{CC}	Supply Voltage	54	4.5	5.0	5.5	V
		74	4.75	5.0	5.25	
T_A	Operating Ambient Temperature Range	54	−55	25	125	°C
		74	0	25	70	
V_{OH}	Output Voltage — High	54, 74			5.5	V
I_{OL}	Output Current — Low	54			4.0	mA
		74			8.0	

DC CHARACTERISTICS OVER OPERATING TEMPERATURE RANGE (unless otherwise specified)

Symbol	Parameter		Min	Typ	Max	Unit	Test Conditions	
				Limits				
			Min	Typ	Max			
V_{IH}	Input HIGH Voltage		2.0			V	Guaranteed Input HIGH Voltage for All Inputs	
V_{IL}	Input LOW Voltage	54			0.7	V	Guaranteed Input LOW Voltage for All Inputs	
		74			0.8			
V_{IK}	Input Clamp Diode Voltage			−0.65	−1.5	V	V_{CC} = MIN, I_{IN} = −18 mA	
I_{OH}	Output HIGH Current	54, 74			100	μA	V_{CC} = MIN, V_{OH} = MAX	
V_{OL}	Output LOW Voltage	54, 74		0.25	0.4	V	I_{OL} = 4.0 mA	V_{CC} = V_{CC} MIN, V_{IN} = V_{IL} or V_{IH} per Truth Table
		74		0.35	0.5	V	I_{OL} = 8.0 mA	
I_{IH}	Input HIGH Current				20	μA	V_{CC} = MAX, V_{IN} = 2.7 V	
					0.1	mA	V_{CC} = MAX, V_{IN} = 7.0 V	
I_{IL}	Input LOW Current				−0.4	mA	V_{CC} = MAX, V_{IN} = 0.4 V	
I_{CC}	Power Supply Current Total, Output HIGH				3.6	mA	V_{CC} = MAX	
	Total, Output LOW				6.6			

AC CHARACTERISTICS (T_A = 25°C)

Symbol	Parameter	Min	Typ	Max	Unit	Test Conditions
			Limits			
		Min	Typ	Max		
t_{PLH}	Turn-Off Delay, Input to Output		20	35	ns	V_{CC} = 5.0 V C_L = 15 pF, R_L = 2.0 kΩ
t_{PHL}	Turn-On Delay, Input to Output		17	35	ns	

 **MOTOROLA**

DUAL 4-INPUT NAND GATE

SN54/74LS20

DUAL 4-INPUT NAND GATE

LOW POWER SCHOTTKY

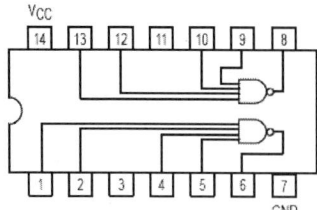

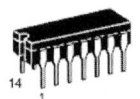

J SUFFIX
CERAMIC
CASE 632-08

N SUFFIX
PLASTIC
CASE 646-06

D SUFFIX
SOIC
CASE 751A-02

ORDERING INFORMATION

SN54LSXXJ   Ceramic
SN74LSXXN Plastic
SN74LSXXD SOIC

GUARANTEED OPERATING RANGES

Symbol	Parameter		Min	Typ	Max	Unit
V_{CC}	Supply Voltage	54	4.5	5.0	5.5	V
		74	4.75	5.0	5.25	
T_A	Operating Ambient Temperature Range	54	−55	25	125	°C
		74	0	25	70	
I_{OH}	Output Current — High	54, 74			−0.4	mA
I_{OL}	Output Current — Low	54			4.0	mA
		74			8.0	

DC CHARACTERISTICS OVER OPERATING TEMPERATURE RANGE (unless otherwise specified)

Symbol	Parameter		Min	Typ	Max	Unit	Test Conditions	
				Limits				
V_{IH}	Input HIGH Voltage		2.0			V	Guaranteed Input HIGH Voltage for All Inputs	
V_{IL}	Input LOW Voltage	54			0.7	V	Guaranteed Input LOW Voltage for All Inputs	
		74			0.8			
V_{IK}	Input Clamp Diode Voltage			−0.65	−1.5	V	V_{CC} = MIN, I_{IN} = −18 mA	
V_{OH}	Output HIGH Voltage	54	2.5	3.5		V	V_{CC} = MIN, I_{OH} = MAX, V_{IN} = V_{IH} or V_{IL} per Truth Table	
		74	2.7	3.5		V		
V_{OL}	Output LOW Voltage	54, 74		0.25	0.4	V	I_{OL} = 4.0 mA	V_{CC} = V_{CC} MIN, V_{IN} = V_{IL} or V_{IH} per Truth Table
		74		0.35	0.5	V	I_{OL} = 8.0 mA	
I_{IH}	Input HIGH Current				20	μA	V_{CC} = MAX, V_{IN} = 2.7 V	
					0.1	mA	V_{CC} = MAX, V_{IN} = 7.0 V	
I_{IL}	Input LOW Current				−0.4	mA	V_{CC} = MAX, V_{IN} = 0.4 V	
I_{OS}	Short Circuit Current (Note 1)		−20		−100	mA	V_{CC} = MAX	
I_{CC}	Power Supply Current Total, Output HIGH				0.8	mA	V_{CC} = MAX	
	Total, Output LOW				2.2			

Note 1: Not more than one output should be shorted at a time, nor for more than 1 second.

AC CHARACTERISTICS (T_A = 25°C)

Symbol	Parameter	Min	Typ	Max	Unit	Test Conditions
			Limits			
t_{PLH}	Turn-Off Delay, Input to Output		9.0	15	ns	V_{CC} = 5.0 V C_L = 15 pF
t_{PHL}	Turn-On Delay, Input to Output		10	15	ns	

SN74LS32

Quad 2-Input OR Gate

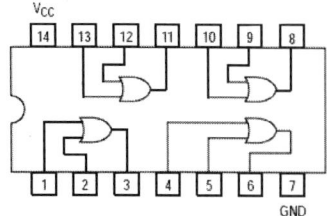

V_CC
14 13 12 11 10 9 8

1 2 3 4 5 6 7
GND

ON Semiconductor
Formerly a Division of Motorola
http://onsemi.com

LOW
POWER
SCHOTTKY

GUARANTEED OPERATING RANGES

Symbol	Parameter	Min	Typ	Max	Unit
V_{CC}	Supply Voltage	4.75	5.0	5.25	V
T_A	Operating Ambient Temperature Range	0	25	70	°C
I_{OH}	Output Current – High			−0.4	mA
I_{OL}	Output Current – Low			8.0	mA

14

1

PLASTIC
N SUFFIX
CASE 646

14

1

SOIC
D SUFFIX
CASE 751A

ORDERING INFORMATION

Device	Package	Shipping
SN74LS32N	14 Pin DIP	2000 Units/Box
SN74LS32D	14 Pin	2500/Tape & Reel

DC CHARACTERISTICS OVER OPERATING TEMPERATURE RANGE (unless otherwise specified)

| Symbol | Parameter | Limits | | | Unit | Test Conditions |
		Min	Typ	Max		
V_{IH}	Input HIGH Voltage	2.0			V	Guaranteed Input HIGH Voltage for All Inputs
V_{IL}	Input LOW Voltage			0.8	V	Guaranteed Input LOW Voltage for All Inputs
V_{IK}	Input Clamp Diode Voltage		−0.65	−1.5	V	V_{CC} = MIN, I_{IN} = −18 mA
V_{OH}	Output HIGH Voltage	2.7	3.5		V	V_{CC} = MIN, I_{OH} = MAX, V_{IN} = V_{IH} or V_{IL} per Truth Table
V_{OL}	Output LOW Voltage		0.25	0.4	V	I_{OL} = 4.0 mA V_{CC} = V_{CC} MIN, V_{IN} = V_{IL} or V_{IH} per Truth Table
			0.35	0.5	V	I_{OL} = 8.0 mA
I_{IH}	Input HIGH Current			20	μA	V_{CC} = MAX, V_{IN} = 2.7 V
				0.1	mA	V_{CC} = MAX, V_{IN} = 7.0 V
I_{IL}	Input LOW Current			−0.4	mA	V_{CC} = MAX, V_{IN} = 0.4 V
I_{OS}	Short Circuit Current (Note 1)	−20		−100	mA	V_{CC} = MAX
I_{CC}	Power Supply Current Total, Output HIGH			6.2	mA	V_{CC} = MAX
	Total, Output LOW			9.8		

Note 1: Not more than one output should be shorted at a time, nor for more than 1 second.

AC CHARACTERISTICS (T_A = 25°C)

| Symbol | Parameter | Limits | | | Unit | Test Conditions |
		Min	Typ	Max		
t_{PLH}	Turn-Off Delay, Input to Output		14	22	ns	V_{CC} = 5.0 V C_L = 15 pF
t_{PHL}	Turn-On Delay, Input to Output		14	22	ns	

SN74LS38

Quad 2-Input NAND Buffer

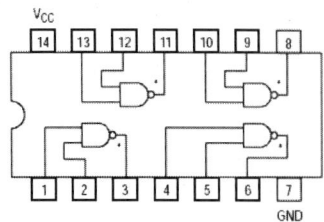

*OPEN COLLECTOR OUTPUTS

ON Semiconductor
Formerly a Division of Motorola
http://onsemi.com

LOW
POWER
SCHOTTKY

PLASTIC
N SUFFIX
CASE 646

SOIC
D SUFFIX
CASE 751A

GUARANTEED OPERATING RANGES

Symbol	Parameter	Min	Typ	Max	Unit
V_{CC}	Supply Voltage	4.75	5.0	5.25	V
T_A	Operating Ambient Temperature Range	0	25	70	°C
V_{OH}	Output Voltage – High			5.5	V
I_{OL}	Output Current – Low			24	mA

ORDERING INFORMATION

Device	Package	Shipping
SN74LS38N	14 Pin DIP	2000 Units/Box
SN74LS38D	14 Pin	2500/Tape & Reel

DC CHARACTERISTICS OVER OPERATING TEMPERATURE RANGE (unless otherwise specified)

Symbol	Parameter	Limits			Unit	Test Conditions	
		Min	Typ	Max			
V_{IH}	Input HIGH Voltage	2.0			V	Guaranteed Input HIGH Voltage for All Inputs	
V_{IL}	Input LOW Voltage			0.8	V	Guaranteed Input LOW Voltage for All Inputs	
V_{IK}	Input Clamp Diode Voltage		−0.65	−1.5	V	V_{CC} = MIN, I_{IN} = −18 mA	
I_{OH}	Output HIGH Current			250	μA	V_{CC} = MIN, V_{OH} = MAX	
V_{OL}	Output LOW Voltage		0.25	0.4	V	I_{OL} = 12 mA	V_{CC} = V_{CC} MIN, V_{IN} = V_{IL} or V_{IH} per Truth Table
			0.35	0.5	V	I_{OL} = 24 mA	
I_{IH}	Input HIGH Current			20	μA	V_{CC} = MAX, V_{IN} = 2.4 V	
				0.1	mA	V_{CC} = MAX, V_{IN} = 7.0 V	
I_{IL}	Input LOW Current			−0.4	mA	V_{CC} = MAX, V_{IN} = 0.4 V	
I_{CC}	Power Supply Current Total, Output HIGH			2.0	mA	V_{CC} = MAX	
	Total, Output LOW			12			

AC CHARACTERISTICS (T_A = 25°C)

Symbol	Parameter	Limits			Unit	Test Conditions
		Min	Typ	Max		
t_{PLH}	Turn-Off Delay, Input to Output		20	32	ns	V_{CC} = 5.0 V, R_L = 667 Ω C_L = 45 pF
t_{PHL}	Turn-On Delay, Input to Output		18	28	ns	

SN74LS47

BCD to 7-Segment Decoder/Driver

The SN74LS47 are Low Power Schottky BCD to 7-Segment Decoder/Drivers consisting of NAND gates, input buffers and seven AND-OR-INVERT gates. They offer active LOW, high sink current outputs for driving indicators directly. Seven NAND gates and one driver are connected in pairs to make BCD data and its complement available to the seven decoding AND-OR-INVERT gates. The remaining NAND gate and three input buffers provide lamp test, blanking input/ripple-blanking output and ripple-blanking input.

The circuits accept 4-bit binary-coded-decimal (BCD) and, depending on the state of the auxiliary inputs, decodes this data to drive a 7-segment display indicator. The relative positive-logic output levels, as well as conditions required at the auxiliary inputs, are shown in the truth tables. Output configurations of the SN74LS47 are designed to withstand the relatively high voltages required for 7-segment indicators.

These outputs will withstand 15 V with a maximum reverse current of 250 μA. Indicator segments requiring up to 24 mA of current may be driven directly from the SN74LS47 high performance output transistors. Display patterns for BCD input counts above nine are unique symbols to authenticate input conditions.

The SN74LS47 incorporates automatic leading and/or trailing-edge zero-blanking control (RBI and RBO). Lamp test (LT) may be performed at any time which the BI/RBO node is a HIGH level. This device also contains an overriding blanking input (BI) which can be used to control the lamp intensity by varying the frequency and duty cycle of the BI input signal or to inhibit the outputs.

- Lamp Intensity Modulation Capability (BI/RBO)
- Open Collector Outputs
- Lamp Test Provision
- Leading/Trailing Zero Suppression
- Input Clamp Diodes Limit High-Speed Termination Effects

ON Semiconductor
Formerly a Division of Motorola
http://onsemi.com

LOW POWER SCHOTTKY

PLASTIC
N SUFFIX
CASE 648

SOIC
D SUFFIX
CASE 751B

GUARANTEED OPERATING RANGES

Symbol	Parameter	Min	Typ	Max	Unit
V_{CC}	Supply Voltage	4.75	5.0	5.25	V
T_A	Operating Ambient Temperature Range	0	25	70	°C
I_{OH}	Output Current – High BI/RBO			−50	μA
I_{OL}	Output Current – Low BI/RBO BI/RBO			3.2	mA
$V_{O(off)}$	Off–State Output Voltage a to g			15	V
$I_{O(on)}$	On–State Output Current a to g			24	mA

ORDERING INFORMATION

Device	Package	Shipping
SN74LS47N	16 Pin DIP	2000 Units/Box
SN74LS47D	16 Pin	2500/Tape & Reel

CONNECTION DIAGRAM DIP (TOP VIEW)

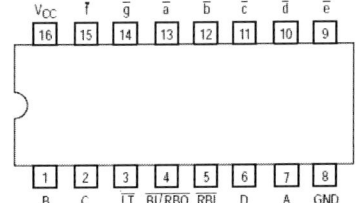

Pin	16	15	14	13	12	11	10	9
	V_{CC}	f	$\bar{g}$	$\bar{a}$	$\bar{b}$	$\bar{c}$	$\bar{d}$	$\bar{e}$

Pin	1	2	3	4	5	6	7	8
	B	C	$\overline{LT}$	$\overline{BI}/\overline{RBO}$	$\overline{RBI}$	D	A	GND

PIN NAMES		LOADING (Note a)	
		HIGH	LOW
A, B, C, D	BCD Inputs	0.5 U.L.	0.25 U.L.
$\overline{RBI}$	Ripple–Blanking Input	0.5 U.L.	0.25 U.L.
$\overline{LT}$	Lamp–Test Input	0.5 U.L.	0.25 U.L.
$\overline{BI}/\overline{RBO}$	Blanking Input or	0.5 U.L.	0.75 U.L.
	Ripple–Blanking Output	1.2 U.L.	2.0 U.L.
$\bar{a}$, to $\bar{g}$	Outputs	Open–Collector	15 U.L.

NOTES:
a) 1 Unit Load (U.L.) = 40 μA HIGH, 1.6 mA LOW.
b) Output current measured at V_{OUT} = 0.5 V
   The Output LOW drive factor is 15 U.L. for Commercial (74) Temperature Ranges.

LOGIC SYMBOL

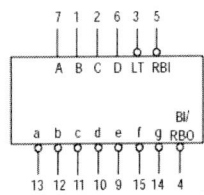

V_{CC} = PIN 16
GND = PIN 8

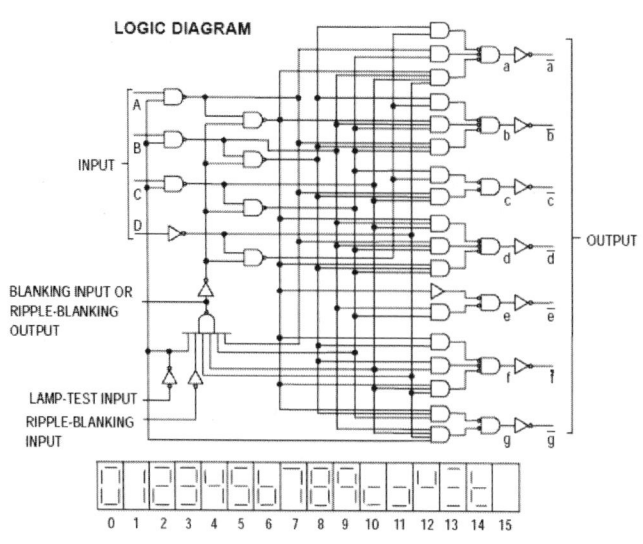

LOGIC DIAGRAM

INPUT — A, B, C, D

BLANKING INPUT OR RIPPLE-BLANKING OUTPUT

LAMP-TEST INPUT

RIPPLE-BLANKING INPUT

OUTPUT — a, ā, b, b̄, c, c̄, d, d̄, e, ē, f, f̄, g, ḡ

```
0  1  2  3  4  5  6  7  8  9  10  11  12  13  14  15
```

NUMERICAL DESIGNATIONS — RESULTANT DISPLAYS

TRUTH TABLE

DECIMAL OR FUNCTION	INPUTS							OUTPUTS							NOTE
	$\overline{LT}$	$\overline{RBI}$	D	C	B	A	$\overline{BI/RBO}$	$\bar{a}$	$\bar{b}$	$\bar{c}$	$\bar{d}$	$\bar{e}$	$\bar{f}$	$\bar{g}$	
0	H	H	L	L	L	L	H	L	L	L	L	L	L	H	A
1	H	X	L	L	L	H	H	H	L	L	H	H	H	H	A
2	H	X	L	L	H	L	H	L	L	H	L	L	H	L	
3	H	X	L	L	H	H	H	L	L	L	L	H	H	L	
4	H	X	L	H	L	L	H	H	L	L	H	H	L	L	
5	H	X	L	H	L	H	H	L	H	L	L	H	L	L	
6	H	X	L	H	H	L	H	H	H	L	L	L	L	L	
7	H	X	L	H	H	H	H	L	L	L	H	H	H	H	
8	H	X	H	L	L	L	H	L	L	L	L	L	L	L	
9	H	X	H	L	L	H	H	L	L	L	H	H	L	L	
10	H	X	H	L	H	L	H	H	H	H	L	L	H	L	
11	H	X	H	L	H	H	H	H	H	L	L	H	H	L	
12	H	X	H	H	L	L	H	H	L	H	H	H	L	L	
13	H	X	H	H	L	H	H	L	H	H	L	H	L	L	
14	H	X	H	H	H	L	H	H	H	H	L	L	L	L	
15	H	X	H	H	H	H	H	H	H	H	H	H	H	H	
$\overline{BI}$	X	X	X	X	X	X	L	H	H	H	H	H	H	H	B
$\overline{RBI}$	H	L	L	L	L	L	L	H	H	H	H	H	H	H	C
$\overline{LT}$	L	X	X	X	X	X	H	L	L	L	L	L	L	L	D

H = HIGH Voltage Level
L = LOW Voltage Level
X = Immaterial

NOTES:
(A) $\overline{BI/RBO}$ is wire-AND logic serving as blanking input (BI) and/or ripple-blanking output ($\overline{RBO}$). The blanking out (BI) must be open or held at a HIGH level when output functions 0 through 15 are desired, and ripple-blanking input ($\overline{RBI}$) must be open or at a HIGH level if blanking of a decimal 0 is not desired. X = input may be HIGH or LOW.
(B) When a LOW level is applied to the blanking input (forced condition) all segment outputs go to a LOW level regardless of the state of any other input condition.
(C) When ripple-blanking input ($\overline{RBI}$) and inputs A, B, C, and D are at LOW level, with the lamp test input at HIGH level, all segment outputs go to a HIGH level and the ripple-blanking output (RBO) goes to a LOW level (response condition).
(D) When the blanking input/ripple-blanking output ($\overline{BI/RBO}$) is open or held at a HIGH level, and a LOW level is applied to lamp test input, all segment outputs go to a LOW level.

DC CHARACTERISTICS OVER OPERATING TEMPERATURE RANGE (unless otherwise specified)

Symbol	Parameter	Min	Typ	Max	Unit	Test Conditions	
V_{IH}	Input HIGH Voltage	2.0			V	Guaranteed Input HIGH Theshold Voltage for All Inputs	
V_{IL}	Input LOW Voltage			0.8	V	Guaranteed Input LOW Threshold Voltage for All Inputs	
V_{IK}	Input Clamp Diode Voltage		−0.65	−1.5	V	V_{CC} = MIN, I_{IN} = − 18 mA	
V_{OH}	Output HIGH Voltage, $\overline{BI}/\overline{RBO}$	2.4	4.2		V	V_{CC} = MIN, I_{OH} = −50 µA, V_{IN} = V_{IN} or V_{IL} per Truth Table	
V_{OL}	Output LOW Voltage $\overline{BI}/\overline{RBO}$		0.25	0.4	V	I_{OL} = 1.6 mA	V_{CC} = MIN, V_{IN} = V_{IN} or V_{IL} per Truth Table
			0.35	0.5	V	I_{OL} = 3.2 mA	
$I_{O\,(off)}$	Off-State Output Current $\overline{a}$ thru $\overline{g}$			250	µA	V_{CC} = MAX, V_{IN} = V_{IN} or V_{IL} per Truth Table, $V_{O\,(off)}$ = 15 V	
$V_{O\,(on)}$	On-State Output Voltage $\overline{a}$ thru $\overline{g}$		0.25	0.4	V	$I_{O\,(on)}$ = 12 mA	V_{CC} = MAX, V_{IN} = V_{IH} or V_{IL} per Truth Table
			0.35	0.5	V	$I_{O\,(on)}$ = 24 mA	
I_{IH}	Input HIGH Current			20	µA	V_{CC} = MAX, V_{IN} = 2.7 V	
				0.1	mA	V_{CC} = MAX, V_{IN} = 7.0 V	
I_{IL}	Input LOW Current $\overline{BI}/\overline{RBO}$ Any Input except $\overline{BI}/\overline{RBO}$			−1.2 −0.4	mA	V_{CC} = MAX, V_{IN} = 0.4 V	
I_{OS} $\overline{BI}/\overline{RBO}$	Output Short Circuit Current (Note 1)	−0.3		−2.0	mA	V_{CC} = MAX, V_{OUT} = 0 V	
I_{CC}	Power Supply Current		7.0	13	mA	V_{CC} = MAX	

Note 1: Not more than one output should be shorted at a time, nor for more than 1 second.

AC CHARACTERISTICS (T_A = 25°C)

Symbol	Parameter	Min	Typ	Max	Unit	Test Conditions
t_{PHL} t_{PLH}	Propagation Delay, Address Input to Segment Output			100 100	ns ns	V_{CC} = 5.0 V C_L = 15 pF
t_{PHL} t_{PLH}	Propagation Delay, $\overline{RBI}$ Input To Segment Output			100 100	ns ns	

AC WAVEFORMS

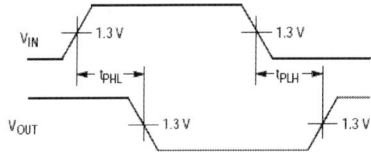

Figure 1.

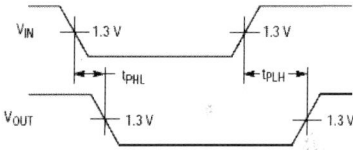

Figure 2.

 **MOTOROLA**

DUAL JK NEGATIVE
EDGE-TRIGGERED FLIP-FLOP

The SN54LS/74LS73A offers individual J, K, clear, and clock inputs. These dual flip-flops are designed so that when the clock goes HIGH, the inputs are enabled and data will be accepted. The logic level of the J and K inputs may be allowed to change when the clock pulse is HIGH and the bistable will perform according to the truth table as long as minimum set-up times are observed. Input data is transferred to the outputs on the negative-going edge of the clock pulse.

LOGIC DIAGRAM (Each Flip-Flop)

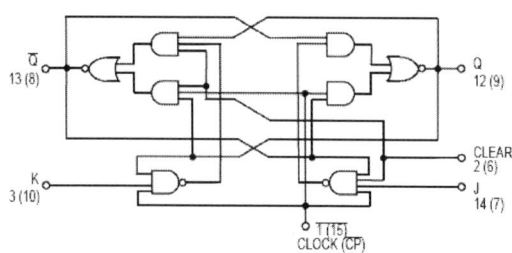

MODE SELECT — TRUTH TABLE

OPERATING MODE	INPUTS			OUTPUTS	
	$\overline{C}_D$	J	K	Q	$\overline{Q}$
Reset (Clear)	L	X	X	L	H
Toggle	H	h	h	$\overline{q}$	q
Load "0" (Reset)	H	l	h	L	H
Load "1" (Set)	H	h	l	H	L
Hold	H	l	l	q	$\overline{q}$

H, h = HIGH Voltage Level
L, l = LOW Voltage Level
X = Don't Care
l, h (q) = Lower case letters indicate the state of the referenced input (or output) one set-up time
    prior to the HIGH to LOW clock transition.

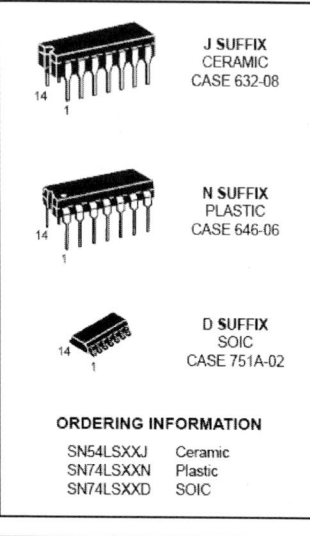
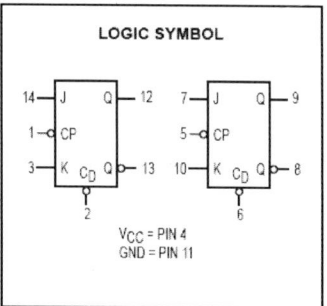

GUARANTEED OPERATING RANGES

Symbol	Parameter		Min	Typ	Max	Unit
V_{CC}	Supply Voltage	54	4.5	5.0	5.5	V
		74	4.75	5.0	5.25	
T_A	Operating Ambient Temperature Range	54	−55	25	125	°C
		74	0	25	70	
I_{OH}	Output Current — High	54, 74			−0.4	mA
I_{OL}	Output Current — Low	54			4.0	mA
		74			8.0	

DC CHARACTERISTICS OVER OPERATING TEMPERATURE RANGE (unless otherwise specified)

Symbol	Parameter		Min	Typ	Max	Unit	Test Conditions	
V_{IH}	Input HIGH Voltage		2.0			V	Guaranteed Input HIGH Voltage for All Inputs	
V_{IL}	Input LOW Voltage	54			0.7	V	Guaranteed Input LOW Voltage for All Inputs	
		74			0.8			
V_{IK}	Input Clamp Diode Voltage			−0.65	−1.5	V	V_{CC} = MIN, I_{IN} = −18 mA	
V_{OH}	Output HIGH Voltage	54	2.5	3.5		V	V_{CC} = MIN, I_{OH} = MAX, V_{IN} = V_{IH} or V_{IL} per Truth Table	
		74	2.7	3.5		V		
V_{OL}	Output LOW Voltage	54, 74		0.25	0.4	V	I_{OL} = 4.0 mA	V_{CC} = V_{CC} MIN, V_{IN} = V_{IL} or V_{IH} per Truth Table
		74		0.35	0.5	V	I_{OL} = 8.0 mA	
I_{IH}	Input HIGH Current	J, K Clear Clock			20 60 80	μA	V_{CC} = MAX, V_{IN} = 2.7 V	
		J, K Clear Clock			0.1 0.3 0.4	mA	V_{CC} = MAX, V_{IN} = 7.0 V	
I_{IL}	Input LOW Current	J, K Clear, Clock			−0.4 −0.8	mA	V_{CC} = MAX, V_{IN} = 0.4 V	
I_{OS}	Short Circuit Current (Note 1)		−20		−100	mA	V_{CC} = MAX	
I_{CC}	Power Supply Current				6.0	mA	V_{CC} = MAX	

Note 1: Not more than one output should be shorted at a time, nor for more than 1 second.

AC CHARACTERISTICS (T_A = 25°C, V_{CC} = 5.0 V)

Symbol	Parameter	Min	Typ	Max	Unit	Test Conditions	
f_{MAX}	Maximum Clock Frequency	30	45		MHz	Figure 1	V_{CC} = 5.0 V C_L = 15 pF
t_{PLH} t_{PHL}	Propagation Delay, Clock to Output		15 15	20 20	ns ns	Figure 1	

AC SETUP REQUIREMENTS (T_A = 25°C)

Symbol	Parameter	Min	Typ	Max	Unit	Test Conditions	
t_W	Clock Pulse Width High	20			ns	Figure 1	V_{CC} = 5.0 V
t_W	Clear Pulse Width	25			ns	Figure 2	
t_S	Setup Time	20			ns	Figure 1	
t_h	Hold Time	0			ns		

SN74LS76A

Dual JK Flip-Flop with Set and Clear

The SN74LS76A offers individual J, K, Clock Pulse, Direct Set and Direct Clear inputs. These dual flip-flops are designed so that when the clock goes HIGH, the inputs are enabled and data will be accepted. The Logic Level of the J and K inputs will perform according to the Truth Table as long as minimum set-up times are observed. Input data is transferred to the outputs on the HIGH-to-LOW clock transitions.

ON Semiconductor
Formerly a Division of Motorola
http://onsemi.com

LOW POWER SCHOTTKY

MODE SELECT – TRUTH TABLE

OPERATING MODE	INPUTS				OUTPUTS	
	$\overline{S}_D$	$\overline{C}_D$	J	K	Q	$\overline{Q}$
Set	L	H	X	X	H	L
Reset (Clear)	H	L	X	X	L	H
*Undetermined	L	L	X	X	H	H
Toggle	H	H	h	h	$\overline{q}$	q
Load "0" (Reset)	H	H	l	h	L	H
Load "1" (Set)	H	H	h	l	H	L
Hold	H	H	l	l	q	$\overline{q}$

* Both outputs will be HIGH while both $\overline{S}_D$ and $\overline{C}_D$ are LOW, but the output states are unpredictable if $\overline{S}_D$ and $\overline{C}_D$ go HIGH simultaneously.

H, h = HIGH Voltage Level

L, l = LOW Voltage Level

X = Immaterial

l, h (q) = Lower case letters indicate the state of the referenced input
(or output) one setup time prior to the HIGH–to–LOW clock transition

PLASTIC
N SUFFIX
CASE 648

SOIC
D SUFFIX
CASE 751B

GUARANTEED OPERATING RANGES

Symbol	Parameter	Min	Typ	Max	Unit
V_{CC}	Supply Voltage	4.75	5.0	5.25	V
T_A	Operating Ambient Temperature Range	0	25	70	°C
I_{OH}	Output Current – High			-0.4	mA
I_{OL}	Output Current – Low			8.0	mA

ORDERING INFORMATION

Device	Package	Shipping
SN74LS76AN	16 Pin DIP	2000 Units/Box
SN74LS76AD	16 Pin	2500/Tape & Reel

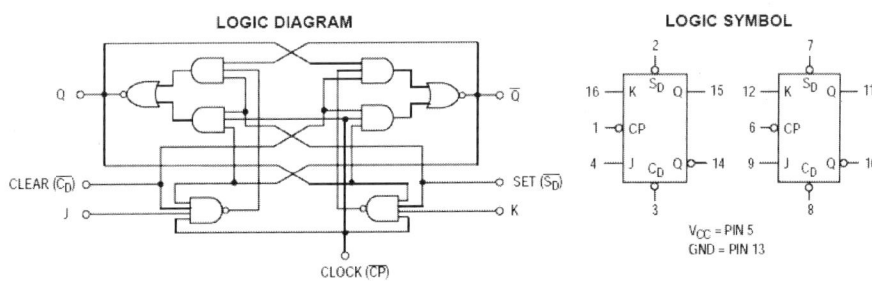

LOGIC DIAGRAM

LOGIC SYMBOL

V_{CC} = PIN 5
GND = PIN 13

DC CHARACTERISTICS OVER OPERATING TEMPERATURE RANGE (unless otherwise specified)

Symbol	Parameter		Min	Typ	Max	Unit	Test Conditions	
				Limits				
V_{IH}	Input HIGH Voltage		2.0			V	Guaranteed Input HIGH Voltage for All Inputs	
V_{IL}	Input LOW Voltage				0.8	V	Guaranteed Input LOW Voltage for All Inputs	
V_{IK}	Input Clamp Diode Voltage			−0.65	−1.5	V	V_{CC} = MIN, I_{IN} = −18 mA	
V_{OH}	Output HIGH Voltage		2.7	3.5		V	V_{CC} = MIN, I_{OH} = MAX, V_{IN} = V_{IH} or V_{IL} per Truth Table	
V_{OL}	Output LOW Voltage			0.25	0.4	V	I_{OL} = 4.0 mA	V_{CC} = V_{CC} MIN, V_{IN} = V_{IL} or V_{IH} per Truth Table
				0.35	0.5	V	I_{OL} = 8.0 mA	
I_{IH}	Input HIGH Current	J, K Clear Clock			20 60 80	μA	V_{CC} = MAX, V_{IN} = 2.7 V	
		J, K Clear Clock			0.1 0.3 0.4	mA	V_{CC} = MAX, V_{IN} = 7.0 V	
I_{IL}	Input LOW Current	J, K Clear, Clock			−0.4 −0.8	mA	V_{CC} = MAX, V_{IN} = 0.4 V	
I_{OS}	Short Circuit Current (Note 1)		−20		−100	mA	V_{CC} = MAX	
I_{CC}	Power Supply Current				6.0	mA	V_{CC} = MAX	

Note 1: Not more than one output should be shorted at a time, nor for more than 1 second.

AC CHARACTERISTICS (T_A = 25°C, V_{CC} = 5.0 V)

Symbol	Parameter	Min	Typ	Max	Unit	Test Conditions
			Limits			
f_{MAX}	Maximum Clock Frequency	30	45		MHz	V_{CC} = 5.0 V, C_L = 15 pF
t_{PLH}	Clock, Clear, Set to Output		15	20	ns	
t_{PHL}			15	20	ns	

AC SETUP REQUIREMENTS (T_A = 25°C)

Symbol	Parameter	Min	Typ	Max	Unit	Test Conditions
			Limits			
t_W	Clock Pulse Width High	20			ns	V_{CC} = 5.0 V
t_W	Clear Set Pulse Width	25			ns	
t_s	Setup Time	20			ns	
t_h	Hold Time	0			ns	

SN74LS86

Quad 2-Input
Exclusive OR Gate

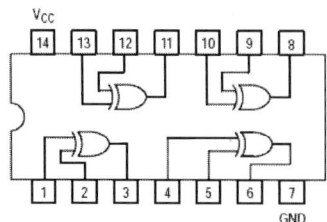

GND

ON Semiconductor
Formerly a Division of Motorola
http://onsemi.com

**LOW
POWER
SCHOTTKY**

TRUTH TABLE

IN		OUT
A	B	Z
L	L	L
L	H	H
H	L	H
H	H	L

PLASTIC
N SUFFIX
CASE 646

GUARANTEED OPERATING RANGES

Symbol	Parameter	Min	Typ	Max	Unit
V_{CC}	Supply Voltage	4.75	5.0	5.25	V
T_A	Operating Ambient Temperature Range	0	25	70	°C
I_{OH}	Output Current – High			−0.4	mA
I_{OL}	Output Current – Low			8.0	mA

SOIC
D SUFFIX
CASE 751A

ORDERING INFORMATION

Device	Package	Shipping
SN74LS86N	14 Pin DIP	2000 Units/Box
SN74LS86D	14 Pin	2500/Tape & Reel

DC CHARACTERISTICS OVER OPERATING TEMPERATURE RANGE (unless otherwise specified)

Symbol	Parameter	Limits			Unit	Test Conditions	
		Min	Typ	Max			
V_{IH}	Input HIGH Voltage	2.0			V	Guaranteed Input HIGH Voltage for All Inputs	
V_{IL}	Input LOW Voltage			0.8	V	Guaranteed Input LOW Voltage for All Inputs	
V_{IK}	Input Clamp Diode Voltage		−0.65	−1.5	V	V_{CC} = MIN, I_{IN} = −18 mA	
V_{OH}	Output HIGH Voltage	2.7	3.5		V	V_{CC} = MIN, I_{OH} = MAX, V_{IN} = V_{IH} or V_{IL} per Truth Table	
V_{OL}	Output LOW Voltage		0.25	0.4	V	I_{OL} = 4.0 mA	V_{CC} = V_{CC} MIN, V_{IN} = V_{IL} or V_{IH} per Truth Table
			0.35	0.5	V	I_{OL} = 8.0 mA	
I_{IH}	Input HIGH Current			40	μA	V_{CC} = MAX, V_{IN} = 2.7 V	
				0.2	mA	V_{CC} = MAX, V_{IN} = 7.0 V	
I_{IL}	Input LOW Current			−0.8	mA	V_{CC} = MAX, V_{IN} = 0.4 V	
I_{OS}	Short Circuit Current (Note 1)	−20		−100	mA	V_{CC} = MAX	
I_{CC}	Power Supply Current			10	mA	V_{CC} = MAX	

Note 1: Not more than one output should be shorted at a time, nor for more than 1 second.

AC CHARACTERISTICS (T_A = 25°C)

Symbol	Parameter	Limits			Unit	Test Conditions
		Min	Typ	Max		
t_{PLH} t_{PHL}	Propagation Delay, Other Input LOW		12 10	23 17	ns	V_{CC} = 5.0 V C_L = 15 pF
t_{PLH} t_{PHL}	Propagation Delay, Other Input HIGH		20 13	30 22	ns	

 **MOTOROLA**

DECADE COUNTER;
DIVIDE-BY-TWELVE COUNTER;
4-BIT BINARY COUNTER

The SN54/74LS90, SN54/74LS92 and SN54/74LS93 are high-speed 4-bit ripple type counters partitioned into two sections. Each counter has a divide-by-two section and either a divide-by-five (LS90), divide-by-six (LS92) or divide-by-eight (LS93) section which are triggered by a HIGH-to-LOW transition on the clock inputs. Each section can be used separately or tied together (Q to $\overline{CP}$) to form BCD, bi-quinary, modulo-12, or modulo-16 counters. All of the counters have a 2-input gated Master Reset (Clear), and the LS90 also has a 2-input gated Master Set (Preset 9).

- Low Power Consumption . . . Typically 45 mW
- High Count Rates . . . Typically 42 MHz
- Choice of Counting Modes . . . BCD, Bi-Quinary, Divide-by-Twelve, Binary
- Input Clamp Diodes Limit High Speed Termination Effects

PIN NAMES

		LOADING (Note a)	
		HIGH	LOW
$\overline{CP}_0$	Clock (Active LOW going edge) Input to ÷2 Section	0.5 U.L.	1.5 U.L.
$\overline{CP}_1$	Clock (Active LOW going edge) Input to ÷5 Section (LS90), ÷6 Section (LS92)	0.5 U.L.	2.0 U.L.
$\overline{CP}_1$	Clock (Active LOW going edge) Input to ÷8 Section (LS93)	0.5 U.L.	1.0 U.L.
MR_1, MR_2	Master Reset (Clear) Inputs	0.5 U.L.	0.25 U.L.
MS_1, MS_2	Master Set (Preset-9, LS90) Inputs	0.5 U.L.	0.25 U.L.
Q_0	Output from ÷2 Section (Notes b & c)	10 U.L.	5 (2.5) U.L.
Q_1, Q_2, Q_3	Outputs from ÷5 (LS90), ÷6 (LS92), ÷8 (LS93) Sections (Note b)	10 U.L.	5 (2.5) U.L.

NOTES:
a. 1 TTL Unit Load (U.L.) = 40 µA HIGH/1.6 mA LOW.
b. The Output LOW drive factor is 2.5 U.L. for Military, (54) and 5 U.L. for commercial (74) Temperature Ranges.
c. The Q_0 Outputs are guaranteed to drive the full fan-out plus the $\overline{CP}_1$ input of the device.
d. To insure proper operation the rise (t_r) and fall time (t_f) of the clock must be less than 100 ns.

SN54/74LS90
SN54/74LS92
SN54/74LS93

DECADE COUNTER;
DIVIDE-BY-TWELVE COUNTER;
4-BIT BINARY COUNTER

LOW POWER SCHOTTKY

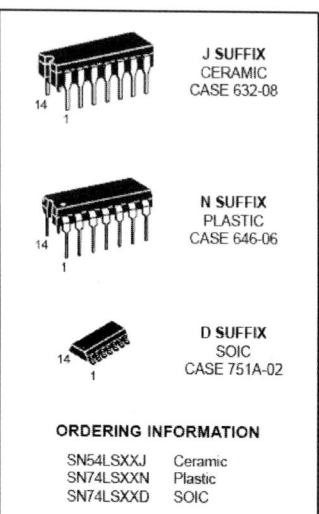

J SUFFIX
CERAMIC
CASE 632-08

N SUFFIX
PLASTIC
CASE 646-06

D SUFFIX
SOIC
CASE 751A-02

ORDERING INFORMATION

SN54LSXXJ	Ceramic
SN74LSXXN	Plastic
SN74LSXXD	SOIC

LOGIC SYMBOL

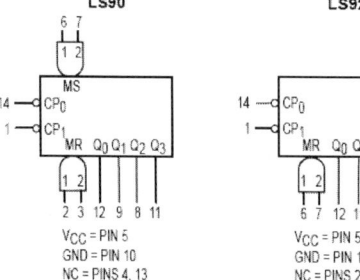

LS90

V_{CC} = PIN 5
GND = PIN 10
NC = PINS 4, 13

LS92

V_{CC} = PIN 5
GND = PIN 10
NC = PINS 2, 3, 4, 13

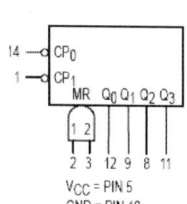

LS93

V_{CC} = PIN 5
GND = PIN 10
NC = PIN 4, 6, 7, 13

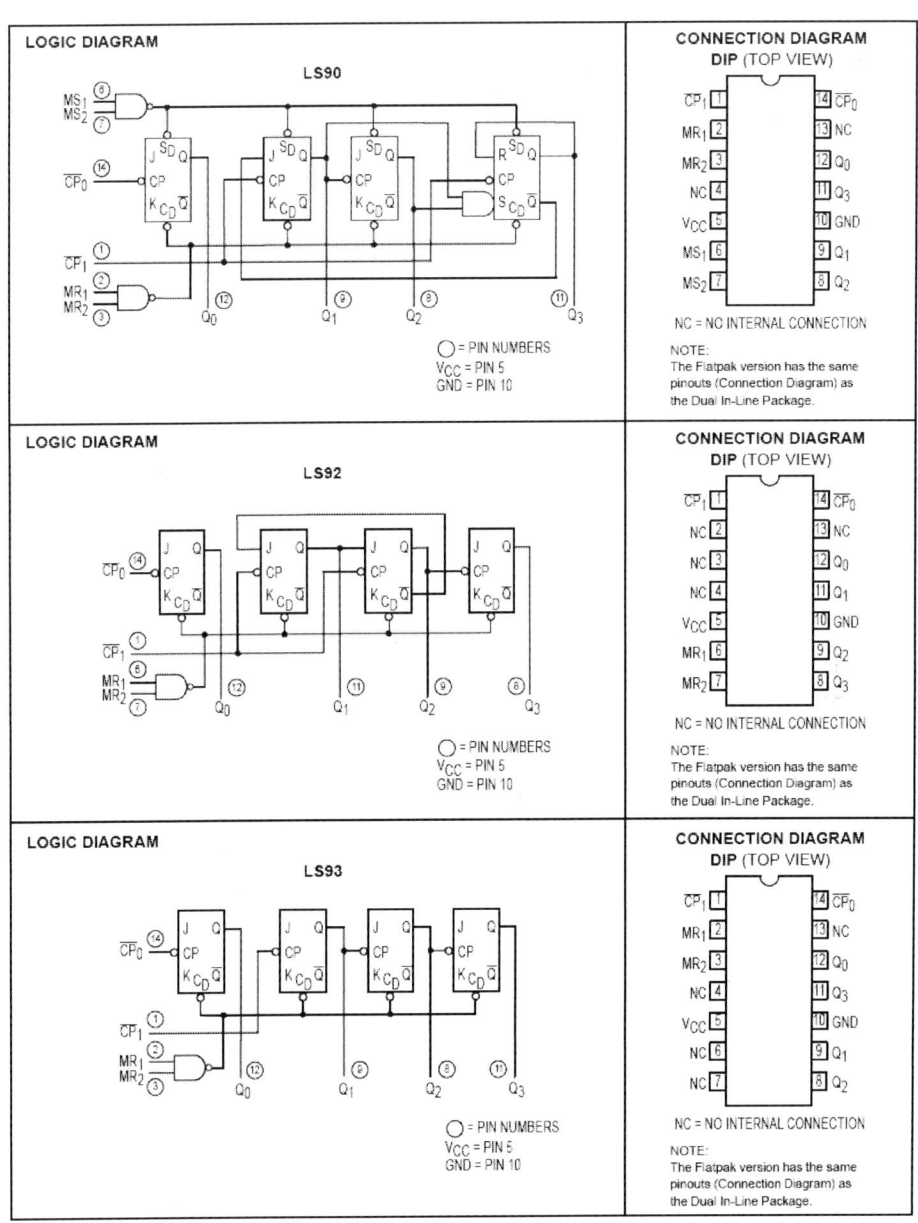

LOGIC DIAGRAM

LS90

MS₁ ⑥
MS₂ ⑦
$\overline{CP}_0$ ⑭
$\overline{CP}_1$ ①
MR₁ ②
MR₂ ③
Q₀ ⑫
Q₁ ⑨
Q₂ ⑧
Q₃ ⑪

○ = PIN NUMBERS
V_CC = PIN 5
GND = PIN 10

CONNECTION DIAGRAM
DIP (TOP VIEW)

$\overline{CP}_1$ 1	14 $\overline{CP}_0$
MR₁ 2	13 NC
MR₂ 3	12 Q₀
NC 4	11 Q₃
V_CC 5	10 GND
MS₁ 6	9 Q₁
MS₂ 7	8 Q₂

NC = NO INTERNAL CONNECTION

NOTE:
The Flatpak version has the same
pinouts (Connection Diagram) as
the Dual In-Line Package.

LOGIC DIAGRAM

LS92

$\overline{CP}_0$ ⑭
$\overline{CP}_1$ ①
⑥
MR₁ ⑦
MR₂
Q₀ ⑫
Q₁ ⑪
Q₂ ⑨
Q₃ ⑧

○ = PIN NUMBERS
V_CC = PIN 5
GND = PIN 10

CONNECTION DIAGRAM
DIP (TOP VIEW)

$\overline{CP}_1$ 1	14 $\overline{CP}_0$
NC 2	13 NC
NC 3	12 Q₀
NC 4	11 Q₁
V_CC 5	10 GND
MR₁ 6	9 Q₂
MR₂ 7	8 Q₃

NC = NO INTERNAL CONNECTION

NOTE:
The Flatpak version has the same
pinouts (Connection Diagram) as
the Dual In-Line Package.

LOGIC DIAGRAM

LS93

$\overline{CP}_0$ ⑭
$\overline{CP}_1$ ①
MR₁ ②
MR₂ ③
Q₀ ⑫
Q₁ ⑨
Q₂ ⑧
Q₃ ⑪

○ = PIN NUMBERS
V_CC = PIN 5
GND = PIN 10

CONNECTION DIAGRAM
DIP (TOP VIEW)

$\overline{CP}_1$ 1	14 $\overline{CP}_0$
MR₁ 2	13 NC
MR₂ 3	12 Q₀
NC 4	11 Q₃
V_CC 5	10 GND
NC 6	9 Q₁
NC 7	8 Q₂

NC = NO INTERNAL CONNECTION

NOTE:
The Flatpak version has the same
pinouts (Connection Diagram) as
the Dual In-Line Package.

SN54/74LS90 • SN54/74LS92 • SN54/74LS93

FUNCTIONAL DESCRIPTION

The LS90, LS92, and LS93 are 4-bit ripple type Decade, Divide-By-Twelve, and Binary Counters respectively. Each device consists of four master/slave flip-flops which are internally connected to provide a divide-by-two section and a divide-by-five (LS90), divide-by-six (LS92), or divide-by-eight (LS93) section. Each section has a separate clock input which initiates state changes of the counter on the HIGH-to-LOW clock transition. State changes of the Q outputs do not occur simultaneously because of internal ripple delays. Therefore, decoded output signals are subject to decoding spikes and should not be used for clocks or strobes. The Q_0 output of each device is designed and specified to drive the rated fan-out plus the $\overline{CP}_1$ input of the device.

A gated AND asynchronous Master Reset ($MR_1 \bullet MR_2$) is provided on all counters which overrides and clocks and resets (clears) all the flip-flops. A gated AND asynchronous Master Set ($MS_1 \bullet MS_2$) is provided on the LS90 which overrides the clocks and the MR inputs and sets the outputs to nine (HLLH).

Since the output from the divide-by-two section is not internally connected to the succeeding stages, the devices may be operated in various counting modes.

LS90

A. BCD Decade (8421) Counter — The $\overline{CP}_1$ input must be externally connected to the Q_0 output. The $\overline{CP}_0$ input receives the incoming count and a BCD count sequence is produced.

B. Symmetrical Bi-quinary Divide-By-Ten Counter — The Q_3 output must be externally connected to the $\overline{CP}_0$ input. The input count is then applied to the $\overline{CP}_1$ input and a divide-by-ten square wave is obtained at output Q_0.

C. Divide-By-Two and Divide-By-Five Counter — No external interconnections are required. The first flip-flop is used as a binary element for the divide-by-two function ($\overline{CP}_0$ as the input and Q_0 as the output). The $\overline{CP}_1$ input is used to obtain binary divide-by-five operation at the Q_3 output.

LS92

A. Modulo 12, Divide-By-Twelve Counter — The $\overline{CP}_1$ input must be externally connected to the Q_0 output. The $\overline{CP}_0$ input receives the incoming count and Q_3 produces a symmetrical divide-by-twelve square wave output.

B. Divide-By-Two and Divide-By-Six Counter —No external interconnections are required. The first flip-flop is used as a binary element for the divide-by-two function. The $\overline{CP}_1$ input is used to obtain divide-by-three operation at the Q_1 and Q_2 outputs and divide-by-six operation at the Q_3 output.

LS93

A. 4-Bit Ripple Counter — The output Q_0 must be externally connected to input $\overline{CP}_1$. The input count pulses are applied to input $\overline{CP}_0$. Simultaneous divisions of 2, 4, 8, and 16 are performed at the Q_0, Q_1, Q_2, and Q_3 outputs as shown in the truth table.

B. 3-Bit Ripple Counter— The input count pulses are applied to input $\overline{CP}_1$. Simultaneous frequency divisions of 2, 4, and 8 are available at the Q_1, Q_2, and Q_3 outputs. Independent use of the first flip-flop is available if the reset function coincides with reset of the 3-bit ripple-through counter.

LS90
MODE SELECTION

RESET/SET INPUTS				OUTPUTS			
MR_1	MR_2	MS_1	MS_2	Q_0	Q_1	Q_2	Q_3
H	H	L	X	L	L	L	L
H	H	X	L	L	L	L	L
X	X	H	H	H	L	L	H
L	X	L	X	Count			
X	L	X	L	Count			
L	X	X	L	Count			
X	L	L	X	Count			

H = HIGH Voltage Level
L = LOW Voltage Level
X = Don't Care

LS92 AND LS93
MODE SELECTION

RESET INPUTS		OUTPUTS			
MR_1	MR_2	Q_0	Q_1	Q_2	Q_3
H	H	L	L	L	L
L	H	Count			
H	L	Count			
L	L	Count			

H = HIGH Voltage Level
L = LOW Voltage Level
X = Don't Care

LS90
BCD COUNT SEQUENCE

COUNT	OUTPUT			
	Q_0	Q_1	Q_2	Q_3
0	L	L	L	L
1	H	L	L	L
2	L	H	L	L
3	H	H	L	L
4	L	L	H	L
5	H	L	H	L
6	L	H	H	L
7	H	H	H	L
8	L	L	L	H
9	H	L	L	H

NOTE: Output Q_0 is connected to Input CP_1 for BCD count.

LS92
TRUTH TABLE

COUNT	OUTPUT			
	Q_0	Q_1	Q_2	Q_3
0	L	L	L	L
1	H	L	L	L
2	L	H	L	L
3	H	H	L	L
4	L	L	H	L
5	H	L	H	L
6	L	L	L	H
7	H	L	L	H
8	L	H	L	H
9	H	H	L	H
10	L	L	H	H
11	H	L	H	H

NOTE: Output Q_0 is connected to Input CP_1.

LS93
TRUTH TABLE

COUNT	OUTPUT			
	Q_0	Q_1	Q_2	Q_3
0	L	L	L	L
1	H	L	L	L
2	L	H	L	L
3	H	H	L	L
4	L	L	H	L
5	H	L	H	L
6	L	H	H	L
7	H	H	H	L
8	L	L	L	H
9	H	L	L	H
10	L	H	L	H
11	H	H	L	H
12	L	L	H	H
13	H	L	H	H
14	L	H	H	H
15	H	H	H	H

NOTE: Output Q_0 is connected to Input CP_1.

GUARANTEED OPERATING RANGES

Symbol	Parameter		Min	Typ	Max	Unit
V_{CC}	Supply Voltage	54	4.5	5.0	5.5	V
		74	4.75	5.0	5.25	
T_A	Operating Ambient Temperature Range	54	−55	25	125	°C
		74	0	25	70	
I_{OH}	Output Current — High	54, 74			−0.4	mA
I_{OL}	Output Current — Low	54			4.0	mA
		74			8.0	

DC CHARACTERISTICS OVER OPERATING TEMPERATURE RANGE (unless otherwise specified)

Symbol	Parameter		Min	Typ	Max	Unit	Test Conditions	
V_{IH}	Input HIGH Voltage		2.0			V	Guaranteed Input HIGH Voltage for All Inputs	
V_{IL}	Input LOW Voltage	54			0.7	V	Guaranteed Input LOW Voltage for All Inputs	
		74			0.8			
V_{IK}	Input Clamp Diode Voltage			−0.65	−1.5	V	V_{CC} = MIN, I_{IN} = −18 mA	
V_{OH}	Output HIGH Voltage	54	2.5	3.5		V	V_{CC} = MIN, I_{OH} = MAX, V_{IN} = V_{IH} or V_{IL} per Truth Table	
		74	2.7	3.5		V		
V_{OL}	Output LOW Voltage	54, 74		0.25	0.4	V	I_{OL} = 4.0 mA	V_{CC} = V_{CC} MIN, V_{IN} = V_{IL} or V_{IH} per Truth Table
		74		0.35	0.5	V	I_{OL} = 8.0 mA	
I_{IH}	Input HIGH Current				20	μA	V_{CC} = MAX, V_{IN} = 2.7 V	
					0.1	mA	V_{CC} = MAX, V_{IN} = 7.0 V	
I_{IL}	Input LOW Current MS, MR $\overline{CP}_0$ CP_1 (LS90, LS92) $\overline{CP}_1$ (LS93)				−0.4 −2.4 −3.2 −1.6	mA	V_{CC} = MAX, V_{IN} = 0.4 V	
I_{OS}	Short Circuit Current (Note 1)		−20		−100	mA	V_{CC} = MAX	
I_{CC}	Power Supply Current				15	mA	V_{CC} = MAX	

Note 1: Not more than one output should be shorted at a time, nor for more than 1 second.

AC CHARACTERISTICS ($T_A = 25°C$, $V_{CC} = 5.0$ V, $C_L = 15$ pF)

Symbol	Parameter	Limits									Unit
		LS90			LS92			LS93			
		Min	Typ	Max	Min	Typ	Max	Min	Typ	Max	
f_{MAX}	$\overline{CP_0}$ Input Clock Frequency	32			32			32			MHz
f_{MAX}	$\overline{CP_1}$ Input Clock Frequency	16			16			16			MHz
t_{PLH} t_{PHL}	Propagation Delay, $\overline{CP_0}$ Input to Q_0 Output		10 12	16 18		10 12	16 18		10 12	16 18	ns
t_{PLH} t_{PHL}	$\overline{CP_0}$ Input to Q_3 Output		32 34	48 50		32 34	48 50		46 46	70 70	ns
t_{PLH} t_{PHL}	$\overline{CP_1}$ Input to Q_1 Output		10 14	16 21		10 14	16 21		10 14	16 21	ns
t_{PLH} t_{PHL}	$\overline{CP_1}$ Input to Q_2 Output		21 23	32 35		10 14	16 21		21 23	32 35	ns
t_{PLH} t_{PHL}	$\overline{CP_1}$ Input to Q_3 Output		21 23	32 35		21 23	32 35		34 34	51 51	ns
t_{PLH}	MS Input to Q_0 and Q_3 Outputs		20	30							ns
t_{PHL}	MS Input to Q_1 and Q_2 Outputs		26	40							ns
t_{PHL}	MR Input to Any Output		26	40		26	40		26	40	ns

AC SETUP REQUIREMENTS ($T_A = 25°C$, $V_{CC} = 5.0$ V)

Symbol	Parameter	Limits						Unit
		LS90		LS92		LS93		
		Min	Max	Min	Max	Min	Max	
t_W	$\overline{CP_0}$ Pulse Width	15		15		15		ns
t_W	$\overline{CP_1}$ Pulse Width	30		30		30		ns
t_W	MS Pulse Width	15						ns
t_W	MR Pulse Width	15		15		15		ns
t_{rec}	Recovery Time MR to $\overline{CP}$	25		25		25		ns

RECOVERY TIME (t_{rec}) is defined as the minimum time required between the end of the reset pulse and the clock transition from HIGH-to-LOW in order to recognize and transfer HIGH data to the Q outputs

AC WAVEFORMS

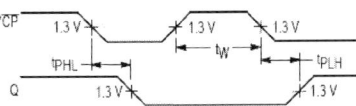

Figure 1

*The number of Clock Pulses required between the t_{PHL} and t_{PLH} measurements can be determined from the appropriate Truth Tables.

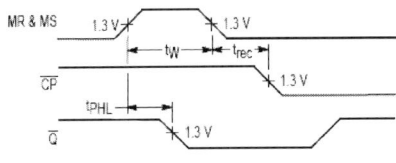

Figure 2

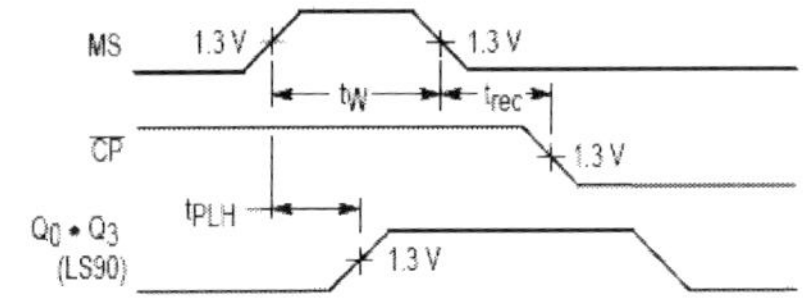

Figure 3

SN74LS122 SN74LS123

Retriggerable Monostable Multivibrators

These dc triggered multivibrators feature pulse width control by three methods. The basic pulse width is programmed by selection of external resistance and capacitance values. The LS122 has an internal timing resistor that allows the circuits to be used with only an external capacitor. Once triggered, the basic pulse width may be extended by retriggering the gated low-level-active (A) or high-level-active (B) inputs, or be reduced by use of the overriding clear.

- Overriding Clear Terminates Output Pulse
- Compensated for V_{CC} and Temperature Variations
- DC Triggered from Active-High or Active-Low Gated Logic Inputs
- Retriggerable for Very Long Output Pulses, up to 100% Duty Cycle
- Internal Timing Resistors on LS122

ON Semiconductor
Formerly a Division of Motorola
http://onsemi.com

LOW POWER SCHOTTKY

PLASTIC
N SUFFIX
CASE 646

SOIC
D SUFFIX
CASE 751A

PLASTIC
N SUFFIX
CASE 648

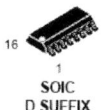

SOIC
D SUFFIX
CASE 751B

GUARANTEED OPERATING RANGES

Symbol	Parameter	Min	Typ	Max	Unit
V_{CC}	Supply Voltage	4.75	5.0	5.25	V
T_A	Operating Ambient Temperature Range	0	25	70	°C
I_{OH}	Output Current – High			−0.4	mA
I_{OL}	Output Current – Low			8.0	mA
R_{ext}	External Timing Resistance	5.0		260	kΩ
C_{ext}	External Capacitance	No Restriction			
R_{ext}/C_{ext}	Wiring Capacitance at R_{ext}/C_{ext} Terminal			50	pF

ORDERING INFORMATION

Device	Package	Shipping
SN74LS122N	14 Pin DIP	2000 Units/Box
SN74LS122D	14 Pin	2500/Tape & Reel
SN74LS123N	16 Pin DIP	2000 Units/Box
SN74LS123D	16 Pin	2500/Tape & Reel

SN74LS122 SN74LS123

SN74LS123 (TOP VIEW)
(SEE NOTES 1 THRU 4)

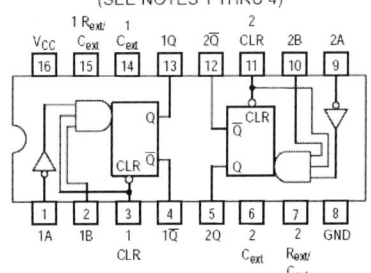

NC — NO INTERNAL CONNECTION.

SN74LS122 (TOP VIEW)
(SEE NOTES 1 THRU 4)

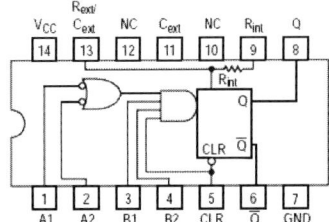

NC — NO INTERNAL CONNECTION.

NOTES:
1. An external timing capacitor may be connected between C_{ext} and R_{ext}/C_{ext} (positive).
2. To use the internal timing resistor of the LS122, connect R_{int} to V_{CC}.
3. For improved pulse width accuracy connect an external resistor between R_{ext}/C_{ext} and V_{CC} with R_{int} open-circuited.
4. To obtain variable pulse widths, connect an external variable resistance between R_{int}/C_{ext} and V_{CC}.

LS122 FUNCTIONAL TABLE

INPUTS					OUTPUTS	
CLEAR	A1	A2	B1	B2	Q	$\overline{Q}$
L	X	X	X	X	L	H
X	H	H	X	X	L	H
X	X	X	L	X	L	H
X	X	X	X	L	L	H
H	L	X	↑	H	⊓	⊔
H	L	X	H	↑	⊓	⊔
H	X	L	↑	H	⊓	⊔
H	X	L	H	↑	⊓	⊔
H	H	↓	H	H	⊓	⊔
H	↓	↓	H	H	⊓	⊔
H	↓	H	H	H	⊓	⊔
↑	L	X	H	H	⊓	⊔
↑	X	L	H	H	⊓	⊔

LS123 FUNCTIONAL TABLE

INPUTS			OUTPUTS	
CLEAR	A	B	Q	Q
L	X	X	L	H
X	H	X	L	H
X	X	L	L	H
H	L	↑	⊓	⊔
H	↓	H	⊓	⊔
↑	L	H	⊓	⊔

TYPICAL APPLICATION DATA

The output pulse t_W is a function of the external components, C_{ext} and R_{ext} or C_{ext} and R_{int} on the LS122. For values of $C_{ext} \geq 1000$ pF, the output pulse at $V_{CC} = 5.0$ V and $V_{RC} = 5.0$ V (see Figures 1, 2, and 3) is given by

$$t_W = K\, R_{ext}\, C_{ext} \text{ where K is nominally } 0.45$$

If C_{ext} is on pF and R_{ext} is in kΩ then t_W is in nanoseconds.

The C_{ext} terminal of the LS122 and LS123 is an internal connection to ground, however for the best system performance C_{ext} should be hard-wired to ground.

Care should be taken to keep R_{ext} and C_{ext} as close to the monostable as possible with a minimum amount of inductance between the R_{ext}/C_{ext} junction and the R_{ext}/C_{ext} pin. Good groundplane and adequate bypassing should be designed into the system for optimum performance to ensure that no false triggering occurs.

It should be noted that the C_{ext} pin is internally connected to ground on the LS122 and LS123, but not on the LS221. Therefore, if C_{ext} is hard-wired externally to ground, substitution of a LS221 onto a LS123 socket will cause the LS221 to become non-functional.

The switching diode is not needed for electrolytic capacitance application and should not be used on the LS122 and LS123.

To find the value of K for $C_{ext} \geq 1000$ pF, refer to Figure 4. Variations on V_{CC} or V_{RC} can cause the value of K to change, as can the temperature of the LS123, LS122.

Figures 5 and 6 show the behavior of the circuit shown in Figures 1 and 2 if separate power supplies are used for V_{CC} and V_{RC}. If V_{CC} is tied to V_{RC}, Figure 7 shows how K will vary with V_{CC} and temperature. Remember, the changes in R_{ext} and C_{ext} with temperature are not calculated and included in the graph.

As long as $C_{ext} \geq 1000$ pF and $5K \leq R_{ext} \leq 260K$, the change in K with respect to R_{ext} is negligible.

If $C_{ext} \leq 1000$ pF the graph shown on Figure 8 can be used to determine the output pulse width. Figure 9 shows how K will change for $C_{ext} \leq 1000$ pF if V_{CC} and V_{RC} are connected to the same power supply. The pulse width t_W in nanoseconds is approximated by

$$t_W = 6 + 0.05\, C_{ext}\,(pF) + 0.45\, R_{ext}\,(k\Omega)\, C_{ext} + 11.6\, R_{ext}$$

In order to trim the output pulse width, it is necessary to include a variable resistor between V_{CC} and the R_{ext}/C_{ext} pin or between V_{CC} and the R_{ext} pin of the LS122. Figure 10, 11, and 12 show how this can be done. R_{ext} remote should be kept as close to the monostable as possible.

Retriggering of the part, as shown in Figure 3, must not occur before C_{ext} is discharged or the retrigger pulse will not have any effect. The discharge time of C_{ext} in nanoseconds is guaranteed to be less than $0.22\, C_{ext}\,(pF)$ and is typically $0.05\, C_{ext}\,(pF)$.

For the smallest possible deviation in output pulse widths from various devices, it is suggested that C_{ext} be kept ≥ 1000 pF.

WAVEFORMS

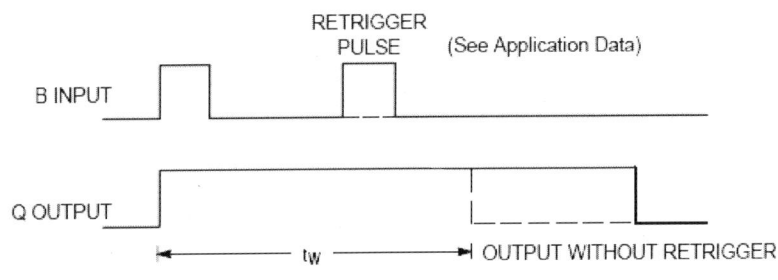

EXTENDING PULSE WIDTH

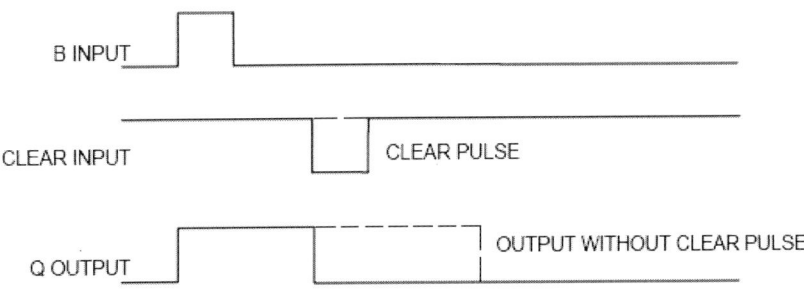

OVERRIDING THE OUTPUT PULSE

DC CHARACTERISTICS OVER OPERATING TEMPERATURE RANGE (unless otherwise specified)

Symbol	Parameter		Limits			Unit	Test Conditions	
			Min	Typ	Max			
V_{IH}	Input HIGH Voltage		2.0			V	Guaranteed Input HIGH Voltage for All Inputs	
V_{IL}	Input LOW Voltage				0.8	V	Guaranteed Input LOW Voltage for All Inputs	
V_{IK}	Input Clamp Diode Voltage			−0.65	−1.5	V	V_{CC} = MIN, I_{IN} = −18 mA	
V_{OH}	Output HIGH Voltage		2.7	3.5		V	V_{CC} = MIN, I_{OH} = MAX, V_{IN} = V_{IH} or V_{IL} per Truth Table	
V_{OL}	Output LOW Voltage			0.25	0.4	V	I_{OL} = 4.0 mA	V_{CC} = V_{CC} MIN, V_{IN} = V_{IL} or V_{IH} per Truth Table
				0.35	0.5	V	I_{OL} = 8.0 mA	
I_{IH}	Input HIGH Current				20	μA	V_{CC} = MAX, V_{IN} = 2.7 V	
					0.1	mA	V_{CC} = MAX, V_{IN} = 7.0 V	
I_{IL}	Input LOW Current				−0.4	mA	V_{CC} = MAX, V_{IN} = 0.4 V	
I_{OS}	Short Circuit Current (Note 1)		−20		−100	mA	V_{CC} = MAX	
I_{CC}	Power Supply Current	LS122			11	mA	V_{CC} = MAX	
		LS123			20			

Note 1: Not more than one output should be shorted at a time, nor for more than 1 second.

AC CHARACTERISTICS (T_A = 25°C, V_{CC} = 5.0 V)

Symbol	Parameter	Limits			Unit	Test Conditions
		Min	Typ	Max		
t_{PLH}	Propagation Delay, A to Q		23	33	ns	
t_{PHL}	Propagation Delay, A to Q̄		32	45		C_{ext} = 0
t_{PLH}	Propagation Delay, B to Q		23	44	ns	C_L = 15 pF
t_{PHL}	Propagation Delay, B to Q̄		34	56		R_{ext} = 5.0 kΩ
t_{PLH}	Propagation Delay, Clear to Q̄		28	45	ns	R_L = 2.0 kΩ
t_{PHL}	Propagation Delay, Clear to Q		20	27		
$t_{W\ min}$	A or B to Q		116	200	ns	C_{ext} = 1000 pF, R_{ext} = 10 kΩ,
$t_W Q$	A to B to Q	4.0	4.5	5.0	μs	C_L = 15 pF, R_L = 2.0 kΩ

AC SETUP REQUIREMENTS (T_A = 25°C, V_{CC} = 5.0 V)

Symbol	Parameter	Limits			Unit	Test Conditions
		Min	Typ	Max		
t_W	Pulse Width	40			ns	

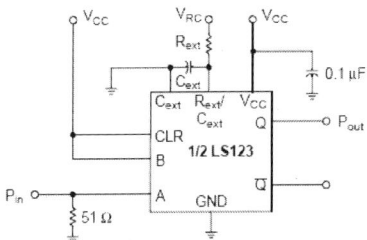

Figure 1.

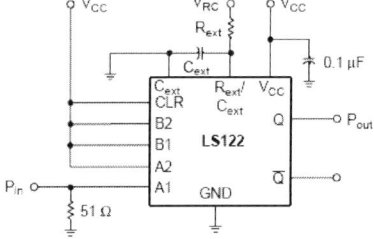

Figure 2.

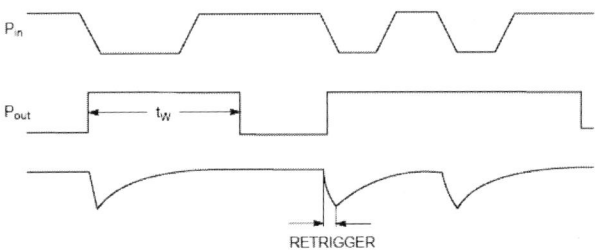

Figure 3.

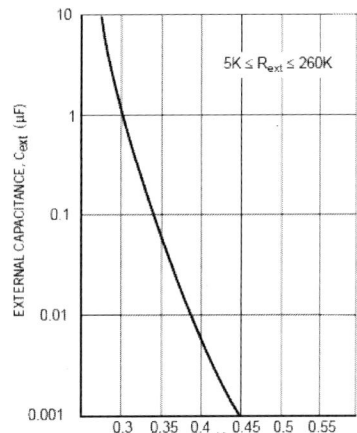

Figure 4.

 **MOTOROLA**

PRESETTABLE BCD/DECADE UP/DOWN COUNTER
PRESETTABLE 4-BIT BINARY UP/DOWN COUNTER

The SN54/74LS192 is an UP/DOWN BCD Decade (8421) Counter and the SN54/74LS193 is an UP/DOWN MODULO-16 Binary Counter. Separate Count Up and Count Down Clocks are used and in either counting mode the circuits operate synchronously. The outputs change state synchronous with the LOW-to-HIGH transitions on the clock inputs.

Separate Terminal Count Up and Terminal Count Down outputs are provided which are used as the clocks for a subsequent stages without extra logic, thus simplifying multistage counter designs. Individual preset inputs allow the circuits to be used as programmable counters. Both the Parallel Load (PL) and the Master Reset (MR) inputs asynchronously override the clocks.

- Low Power . . . 95 mW Typical Dissipation
- High Speed . . . 40 MHz Typical Count Frequency
- Synchronous Counting
- Asynchronous Master Reset and Parallel Load
- Individual Preset Inputs
- Cascading Circuitry Internally Provided
- Input Clamp Diodes Limit High Speed Termination Effects

CONNECTION DIAGRAM DIP (TOP VIEW)

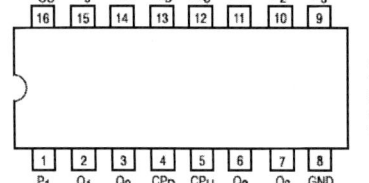

NOTE:
The Flatpak version has the same pinouts (Connection Diagram) as the Dual In-Line Package.

PIN NAMES

		LOADING (Note a)	
		HIGH	LOW
CP_U	Count Up Clock Pulse Input	0.5 U.L.	0.25 U.L.
CP_D	Count Down Clock Pulse Input	0.5 U.L.	0.25 U.L.
$\overline{MR}$	Asynchronous Master Reset (Clear) Input	0.5 U.L.	0.25 U.L.
$\overline{PL}$	Asynchronous Parallel Load (Active LOW) Input	0.5 U.L.	0.25 U.L.
P_n	Parallel Data Inputs	0.5 U.L.	0.25 U.L.
Q_n	Flip-Flop Outputs (Note b)	10 U.L.	5 (2.5) U.L.
$\overline{TC_D}$	Terminal Count Down (Borrow) Output (Note b)	10 U.L.	5 (2.5) U.L.
TC_U	Terminal Count Up (Carry) Output (Note b)	10 U.L.	5 (2.5) U.L.

NOTES:
a. 1 TTL Unit Load (U.L.) = 40 μA HIGH/1.6 mA LOW.
b. The Output LOW drive factor is 2.5 U.L. for Military (54) and 5 U.L. for Commercial (74) Temperature Ranges.

SN54/74LS192
SN54/74LS193

PRESETTABLE BCD/DECADE UP/DOWN COUNTER

PRESETTABLE 4-BIT BINARY UP/DOWN COUNTER

LOW POWER SCHOTTKY

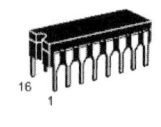

J SUFFIX
CERAMIC
CASE 620-09

N SUFFIX
PLASTIC
CASE 648-08

D SUFFIX
SOIC
CASE 751B-03

ORDERING INFORMATION

SN54LSXXXJ	Ceramic
SN74LSXXXN	Plastic
SN74LSXXXD	SOIC

LOGIC SYMBOL

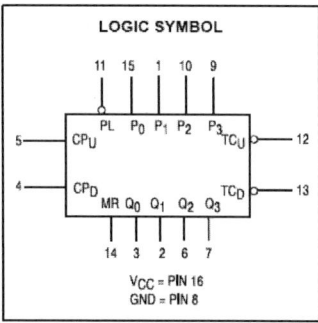

V_{CC} = PIN 16
GND = PIN 8

STATE DIAGRAMS

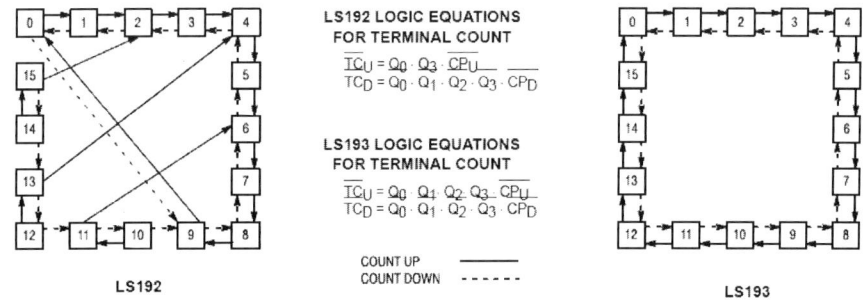

LS192 LOGIC EQUATIONS FOR TERMINAL COUNT

$$\overline{TC_U} = Q_0 \cdot Q_3 \cdot \overline{CP_U}$$
$$TC_D = Q_0 \cdot Q_1 \cdot Q_2 \cdot Q_3 \cdot CP_D$$

LS193 LOGIC EQUATIONS FOR TERMINAL COUNT

$$\overline{TC_U} = Q_0 \cdot Q_1 \cdot Q_2 \cdot Q_3 \cdot \overline{CP_U}$$
$$TC_D = Q_0 \cdot Q_1 \cdot Q_2 \cdot Q_3 \cdot CP_D$$

COUNT UP ———————
COUNT DOWN - - - - - -

LS192   LS193

LOGIC DIAGRAMS

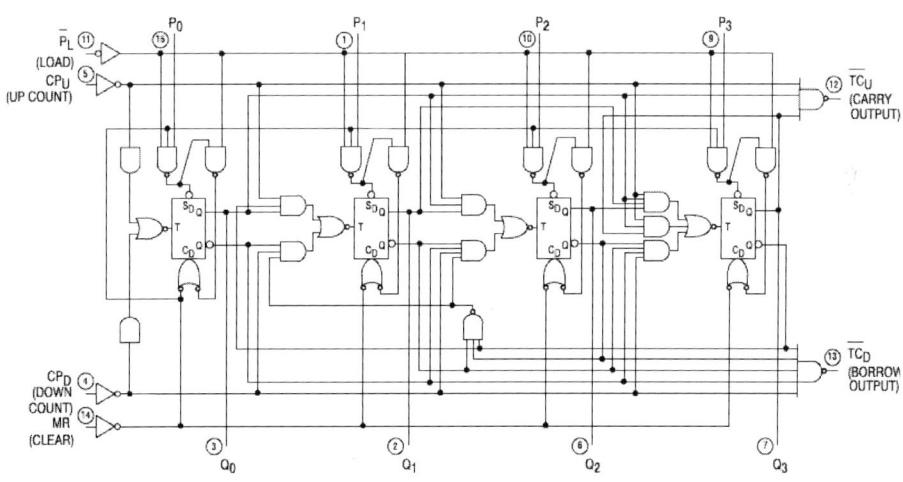

LS192

V_{CC} = PIN 16
GND = PIN 8
○ = PIN NUMBERS

LOGIC DIAGRAMS (continued)

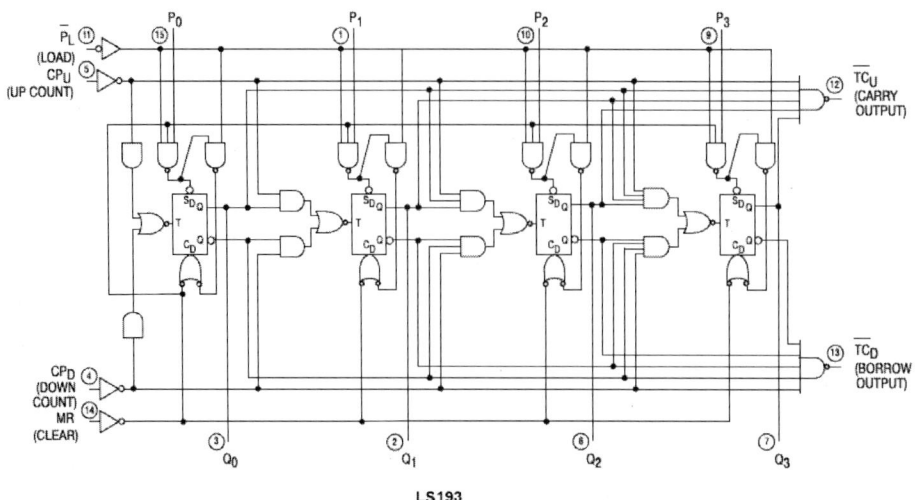

LS193

V_{CC} = PIN 16
GND = PIN 8
○ = PIN NUMBERS

FUNCTIONAL DESCRIPTION

The LS192 and LS193 are Asynchronously Presettable Decade and 4-Bit Binary Synchronous UP/DOWN (Reversable) Counters. The operating modes of the LS192 decade counter and the LS193 binary counter are identical, with the only difference being the count sequences as noted in the State Diagrams. Each circuit contains four master/slave flip-flops, with internal gating and steering logic to provide master reset, individual preset, count up and count down operations.

Each flip-flop contains JK feedback from slave to master such that a LOW-to-HIGH transition on its T input causes the slave, and thus the Q output to change state. Synchronous switching, as opposed to ripple counting, is achieved by driving the steering gates of all stages from a common Count Up line and a common Count Down line, thereby causing all state changes to be initiated simultaneously. A LOW-to-HIGH transition on the Count Up input will advance the count by one; a similar transition on the Count Down input will decrease the count by one. While counting with one clock input, the other should be held HIGH. Otherwise, the circuit will either count by twos or not at all, depending on the state of the first flip-flop, which cannot toggle as long as either Clock input is LOW.

The Terminal Count Up ($\overline{TC_U}$) and Terminal Count Down ($\overline{TC_D}$) outputs are normally HIGH. When a circuit has reached the maximum count state (9 for the LS192, 15 for the LS193), the next HIGH-to-LOW transition of the Count Up Clock will cause TC_U to go LOW. TC_U will stay LOW until CP_U goes HIGH again, thus effectively repeating the Count Up Clock, but delayed by two gate delays. Similarly, the TC_D output will go LOW when the circuit is in the zero state and the Count Down Clock goes LOW. Since the TC outputs repeat the clock waveforms, they can be used as the clock input signals to the next higher order circuit in a multistage counter.

Each circuit has an asynchronous parallel load capability permitting the counter to be preset. When the Parallel Load (PL) and the Master Reset (MR) inputs are LOW, information present on the Parallel Data inputs (P_0, P_3) is loaded into the counter and appears on the outputs regardless of the conditions of the clock inputs. A HIGH signal on the Master Reset input will disable the preset gates, override both Clock inputs, and latch each Q output in the LOW state. If one of the Clock inputs is LOW during and after a reset or load operation, the next LOW-to-HIGH transition of that Clock will be interpreted as a legitimate signal and will be counted.

MODE SELECT TABLE

MR	PL	CP_U	CP_D	MODE
H	X	X	X	Reset (Asyn.)
L	L	X	X	Preset (Asyn.)
L	H	H	H	No Change
L	H	⌐	H	Count Up
L	H	H	⌐	Count Down

L = LOW Voltage Level
H = HIGH Voltage Level
X = Don't Care
⌐ = LOW-to-HIGH Clock Transition

GUARANTEED OPERATING RANGES

Symbol	Parameter		Min	Typ	Max	Unit
V_{CC}	Supply Voltage	54	4.5	5.0	5.5	V
		74	4.75	5.0	5.25	
T_A	Operating Ambient Temperature Range	54	−55	25	125	°C
		74	0	25	70	
I_{OH}	Output Current — High	54, 74			−0.4	mA
I_{OL}	Output Current — Low	54			4.0	mA
		74			8.0	

DC CHARACTERISTICS OVER OPERATING TEMPERATURE RANGE (unless otherwise specified)

Symbol	Parameter		Limits			Unit	Test Conditions	
			Min	Typ	Max			
V_{IH}	Input HIGH Voltage		2.0			V	Guaranteed Input HIGH Voltage for All Inputs	
V_{IL}	Input LOW Voltage	54			0.7	V	Guaranteed Input LOW Voltage for All Inputs	
		74			0.8			
V_{IK}	Input Clamp Diode Voltage			−0.65	−1.5	V	V_{CC} = MIN, I_{IN} = −18 mA	
V_{OH}	Output HIGH Voltage	54	2.5	3.5		V	V_{CC} = MIN, I_{OH} = MAX, V_{IN} = V_{IH} or V_{IL} per Truth Table	
		74	2.7	3.5		V		
V_{OL}	Output LOW Voltage	54, 74		0.25	0.4	V	I_{OL} = 4.0 mA	V_{CC} = V_{CC} MIN, V_{IN} = V_{IL} or V_{IH} per Truth Table
		74		0.35	0.5	V	I_{OL} = 8.0 mA	
I_{IH}	Input HIGH Current				20	µA	V_{CC} = MAX, V_{IN} = 2.7 V	
					0.1	mA	V_{CC} = MAX, V_{IN} = 7.0 V	
I_{IL}	Input LOW Current				−0.4	mA	V_{CC} = MAX, V_{IN} = 0.4 V	
I_{OS}	Short Circuit Current (Note 1)		−20		−100	mA	V_{CC} = MAX	
I_{CC}	Power Supply Current				34	mA	V_{CC} = MAX	

Note 1: Not more than one output should be shorted at a time, nor for more than 1 second.

AC CHARACTERISTICS (T_A = 25°C)

Symbol	Parameter	Limits			Unit	Test Conditions
		Min	Typ	Max		
f_{MAX}	Maximum Clock Frequency	25	32		MHz	
t_{PLH} t_{PHL}	CP_U Input to TC_U Output		17 18	26 24	ns	
t_{PLH} t_{PHL}	CP_D Input to TC_D Output		16 15	24 24	ns	V_{CC} = 5.0 V C_L = 15 pF
t_{PLH} t_{PHL}	Clock to Q		27 30	38 47	ns	
t_{PLH} t_{PHL}	$\overline{PL}$ to Q		24 25	40 40	ns	
t_{PHL}	MR Input to Any Output		23	35	ns	

AC SETUP REQUIREMENTS ($T_A = 25°C$)

Symbol	Parameter	Limits			Unit	Test Conditions
		Min	Typ	Max		
t_W	Any Pulse Width	20			ns	
t_S	Data Setup Time	20			ns	$V_{CC} = 5.0$ V
t_h	Data Hold Time	5.0			ns	
t_{rec}	Recovery Time	40			ns	

DEFINITIONS OF TERMS

SETUP TIME (t_S) is defined as the minimum time required for the correct logic level to be present at the logic input prior to the PL transition from LOW-to-HIGH in order to be recognized and transferred to the outputs.

HOLD TIME (t_h) is defined as the minimum time following the PL transition from LOW-to-HIGH that the logic level must be maintained at the input in order to ensure continued recogni-tion. A negative HOLD TIME indicates that the correct logic level may be released prior to the PL transition from LOW-to-HIGH and still be recognized.

RECOVERY TIME (t_{rec}) is defined as the minimum time required between the end of the reset pulse and the clock transition from LOW-to-HIGH in order to recognize and transfer HIGH data to the Q outputs.

MOTOROLA

QUAD 2-INPUT
EXCLUSIVE NOR GATE

SN54/74LS266

QUAD 2-INPUT
EXCLUSIVE NOR GATE

LOW POWER SCHOTTKY

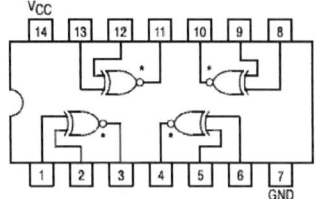

*OPEN COLLECTOR OUTPUTS

TRUTH TABLE

IN		OUT
A	B	Z
L	L	H
L	H	L
H	L	L
H	H	H

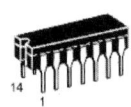

J SUFFIX
CERAMIC
CASE 632-08

N SUFFIX
PLASTIC
CASE 646-06

D SUFFIX
SOIC
CASE 751A-02

ORDERING INFORMATION

SN54LSXXXJ   Ceramic
SN74LSXXXN Plastic
SN74LSXXXD SOIC

GUARANTEED OPERATING RANGES

Symbol	Parameter		Min	Typ	Max	Unit
V_{CC}	Supply Voltage	54	4.5	5.0	5.5	V
		74	4.75	5.0	5.25	
T_A	Operating Ambient Temperature Range	54	−55	25	125	°C
		74	0	25	70	
V_{OH}	Output Voltage — High	54, 74			5.5	V
I_{OL}	Output Current — Low	54			4.0	mA
		74			8.0	

DC CHARACTERISTICS OVER OPERATING TEMPERATURE RANGE (unless otherwise specified)

Symbol	Parameter		Min	Typ	Max	Unit	Test Conditions	
V_{IH}	Input HIGH Voltage		2.0			V	Guaranteed Input HIGH Voltage for All Inputs	
V_{IL}	Input LOW Voltage	54			0.7	V	Guaranteed Input LOW Voltage for All Inputs	
		74			0.8			
V_{IK}	Input Clamp Diode Voltage			−0.65	−1.5	V	V_{CC} = MIN, I_{IN} = −18 mA	
V_{OH}	Output HIGH Voltage	54, 74			100	μA	V_{CC} = MIN, V_{OH} = MAX	
V_{OL}	Output LOW Voltage	54, 74		0.25	0.4	V	I_{OL} = 4.0 mA	V_{CC} = V_{CC} MIN, V_{IN} = V_{IL} or V_{IH} per Truth Table
		74		0.35	0.5	V	I_{OL} = 8.0 mA	
I_{IH}	Input HIGH Current				40	μA	V_{CC} = MAX, V_{IN} = 2.7 V	
					0.2	mA	V_{CC} = MAX, V_{IN} = 7.0 V	
I_{IL}	Input LOW Current				−0.8	mA	V_{CC} = MAX, V_{IN} = 0.4 V	
I_{CC}	Power Supply Current				13	mA	V_{CC} = MAX	

AC CHARACTERISTICS (T_A = 25°C)

Symbol	Parameter	Min	Typ	Max	Unit	Test Conditions
t_{PLH} t_{PHL}	Propagation Delay, Other Input LOW		18 18	30 30	ns	V_{CC} = 5.0 V C_L = 15 pF, R_L = 2.0 kΩ
t_{PLH} t_{PHL}	Propagation Delay, Other Input HIGH		18 18	30 30	ns	

Data Sheet   September 1998   File Number   976.3

4MHz, BiMOS Operational Amplifier with MOSFET Input/CMOS Output

The CA3160A and CA3160 are integrated circuit operational amplifiers that combine the advantage of both CMOS and bipolar transistors on a monolithic chip. The CA3160 series are frequency compensated versions of the popular CA3130 series.

Gate protected P-Channel MOSFET (PMOS) transistors are used in the input circuit to provide very high input impedance, very low input current, and exceptional speed performance. The use of PMOS field effect transistors in the input stage results in common-mode input voltage capability down to 0.5V below the negative supply terminal, an important attribute in single supply applications.

A complementary symmetry MOS (CMOS) transistor-pair, capable of swinging the output voltage to within 10mV of either supply voltage terminal (at very high values of load impedance), is employed as the output circuit.

The CA3160 Series circuits operate at supply voltages ranging from 5V to 16V, or ±2.5V to ±8V when using split supplies, and have terminals for adjustment of offset voltage for applications requiring offset null capability. Terminal provisions are also made to permit strobing of the output stage.

The CA3160A offers superior input characteristics over those of the CA3160.

Ordering Information

PART NUMBER	TEMP. RANGE (°C)	PACKAGE	PKG. NO.
CA3160AE	-55 to 125	8 Ld PDIP	E8.3
CA3160E	-55 to 125	8 Ld PDIP	E8.3
CA3160T	-55 to 125	8 Pin Metal Can	T8.C

Features

- MOSFET Input Stage Provides:
  - Very High Z_I = 1.5TΩ (1.5 x $10^{12}\Omega$) (Typ)
  - Very Low I_I . . . . 5pA (Typ) at 15V Operation
    . . . . 2pA (Typ) at 5V Operation
- Common-Mode Input Voltage Range Includes Negative Supply Rail; Input Terminals Can Be Swung 0.5V Below Negative Supply Rail
- CMOS Output Stage Permits Signal Swing to Either (or Both) Supply Rails

Applications

- Ground Referenced Single Supply Amplifiers
- Fast Sample Hold Amplifiers
- Long Duration Timers/Monostables
- High Input Impedance Wideband Amplifiers
- Voltage Followers (e.g., Follower for Single Supply D/A Converter)
- Wien-Bridge Oscillators
- Voltage Controlled Oscillators
- Photo Diode Sensor Amplifiers

Pinouts

CA3160
(METAL CAN)
TOP VIEW

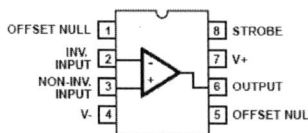

SUPPLEMENTARY
COMPENSATION   TAB   STROBE

OFFSET NULL (1)   (7) V+

INV. INPUT (2)   (6) OUTPUT

NON-INV. INPUT (3)   (5) OFFSET NULL

(4)

V- AND CASE

CA3160
(PDIP)
TOP VIEW

OFFSET NULL [1]   [8] STROBE

INV. INPUT [2]   [7] V+

NON-INV. INPUT [3]   [6] OUTPUT

V- [4]   [5] OFFSET NULL

NOTE: CA3160 Series devices have an on-chip frequency compensation network. Supplementary phase compensation or frequency roll-off (if desired) can be connected externally between Terminals 1 and 8.

Absolute Maximum Ratings

Supply Voltage (Between V+ and V- Terminals) . . . . +16V
Differential Mode Input Voltage 8V
Input Voltage . . . . (V+ +8V) to (V- -0.5V)
Input Current 1mA
Output Short Circuit Duration (Note 2). . . . . Indefinite

Operating Conditions

Temperature Range . . . . -55°C to 125°C

Thermal Information

Thermal Resistance (Typical, Note 1)	θ_{JA} (°C/W)	θ_{JC} (°C/W)
PDIP Package	110	N/A
Metal Can Package	170	85

Maximum Junction Temperature (Metal Can). . . . . .175°C
Maximum Junction Temperature (Plastic Package) . . . . .150°C
Maximum Storage Temperature Range . . . . -65°C to 150°C
Maximum Lead Temperature (Soldering 10s) . . . . 300°C

CAUTION: Stresses above those listed in "Absolute Maximum Ratings" may cause permanent damage to the device. This is a stress only rating and operation of the device at these or any other conditions above those indicated in the operational sections of this specification is not implied.

NOTES:

1. θ_{JA} is measured with the component mounted on an evaluation PC board in free air.

2. Short Circuit may be applied to ground or to either supply.

Electrical Specifications T_A = 25°C, V+ = 15V, V- = 0V, Unless Otherwise Specified

PARAMETER	SYMBOL	TEST CONDITIONS	CA3160			CA3160A			UNITS
			MIN	TYP	MAX	MIN	TYP	MAX	
Input Offset Voltage	$|V_{IO}|$	$V_S = \pm7.5V$	-	6	15	-	2	5	mV
Input Offset Current	$|I_{IO}|$	$V_S = \pm7.5V$	-	0.5	30	-	0.5	20	pA
Input Current	I_I	$V_S = \pm7.5V$	-	5	50	-	5	30	pA
Large-Signal Voltage Gain	A_{OL}	$V_O = 10V_{P-P}$, $R_L = 2k\Omega$	50	320	-	50	320	-	kV/V
			94	110	-	94	110	-	dB
Common-Mode Rejection Ratio	CMRR		70	90	-	80	95	-	dB
Common-Mode Input-Voltage Range	V_{ICR}		0	-0.5 to 12	10	0	-0.5 to 12	10	V
Power-Supply Rejection Ratio	PSRR	$\Delta V_{IO}/\Delta V_S$, $V_S = \pm7.5V$	-	32	320	-	32	150	µV/V
Maximum Output Voltage	$V_{OM}+$	$R_L = 2k\Omega$	12	13.3	-	12	13.3	-	V
	$V_{OM}-$		-	0.002	0.01	-	0.002	0.01	V
	$V_{OM}+$	$R_L = \infty$	14.99	15	-	14.99	15	-	V
	$V_{OM}-$		-	0	0.01	-	0	0.01	V
Maximum Output Current	$I_{OM}+$	$V_O = 0V$ (Source)	12	22	45	12	22	45	mA
	$I_{OM}-$	$V_O = 15V$ (Sink)	12	20	45	12	20	45	mA
Supply Current (Note 3)	I+	$V_O = 7.5V$, $R_L = \infty$	-	10	15	-	10	15	mA
		$V_O = 0V$, $R_L = \infty$	-	2	3	-	2	3	mA
Input Offset Voltage Temperature Drift		$\Delta V_{IO}/\Delta T$	-	8	-	-	6	-	µV/°C

Electrical Specifications For Design Guidance, $V_{SUPPLY} = \pm7.5V$, T_A = 25°C, Unless Otherwise Specified

PARAMETER	SYMBOL	TEST CONDITIONS		CA3160 TYP	CA3160A TYP	UNITS
Input Offset Voltage Adjustment Range		10kΩ Across Terminals 4 and 5 or Terminals 4 and 1		±22	±22	mV
Input Resistance	R_I			1.5	1.5	TΩ
Input Capacitance	C_I	f = 1MHz		4.3	4.3	pF
Equivalent Input Noise Voltage	e_N	BW = 0.2MHz	$R_S = 1M\Omega$	40	40	µV
			$R_S = 10M\Omega$	50	50	µV
Equivalent Input Noise Voltage	e_N	$R_S = 100\Omega$	1kHz	72	72	nV/√Hz
			10kHz	30	30	nV/√Hz

Electrical Specifications For Design Guidance, $V_{SUPPLY} = \pm 7.5V$, $T_A = 25^\circ C$, Unless Otherwise Specified **(Continued)**

PARAMETER		SYMBOL	TEST CONDITIONS	CA3160 TYP	CA3160A TYP	UNITS
Unity Gain Crossover Frequency		f_T		4	4	MHz
Slew Rate		SR		10	10	V/µs
Transient Response	Rise and Fall Time	t_r	$C_L = 25pF$, $R_L = 2k\Omega$, (Voltage Follower)	0.09	0.09	µs
	Overshoot	OS		10	10	%
Settling Time		t_S	$C_L = 25pF$, $R_L = 2k\Omega$, (Voltage Follower) To <0.1%, $V_{IN} = 4V_{P-P}$	1.8	1.8	µs

Electrical Specifications For Design Guidance, V+ = +5V, V- = 0V, $T_A = 25^\circ C$, Unless Otherwise Specified

PARAMETER	SYMBOL	TEST CONDITIONS	CA3160 TYP	CA3160A TYP	UNITS
Input Offset Voltage	V_{IO}		6	2	mV
Input Offset Current	I_{IO}		0.1	0.1	pA
Input Current	I_I		2	2	pA
Common-Mode Rejection Ratio	CMRR		80	90	dB
Large Signal Voltage Gain	A_{OL}	$V_O = 4V_{P-P}$, $R_L = 5k\Omega$	100	100	kV/V
			100	100	dB
Common-Mode Input Voltage Range	V_{ICR}		0 to 2.8	0 to 2.8	V
Supply Current	I+	$V_O = 5V$, $R_L = \infty$	300	300	µA
		$V_O = 2.5V$, $R_L = \infty$	500	500	µA
Power Supply Rejection Ratio	PSRR	$\Delta V_{IO}/\Delta V+$	200	200	µV/V

NOTE:

3. I_{CC} typically increases by 1.5mA/MHz during operation.

Block Diagram

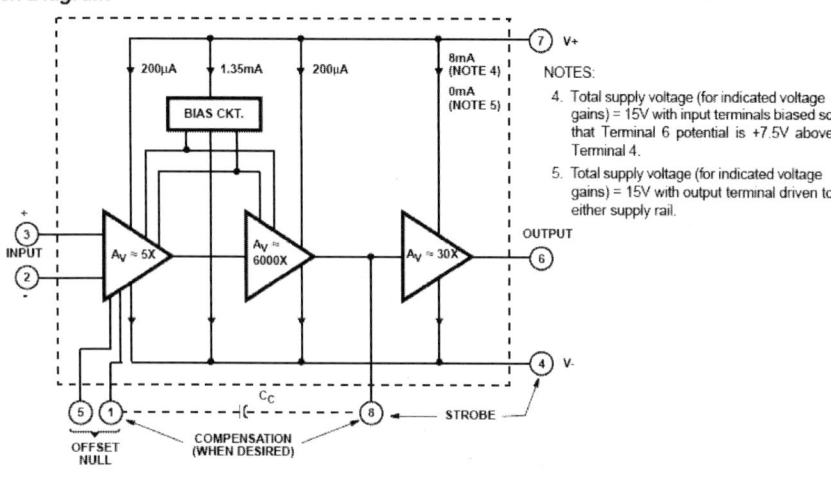

NOTES:

4. Total supply voltage (for indicated voltage gains) = 15V with input terminals biased so that Terminal 6 potential is +7.5V above Terminal 4.

5. Total supply voltage (for indicated voltage gains) = 15V with output terminal driven to either supply rail.

Schematic Diagram

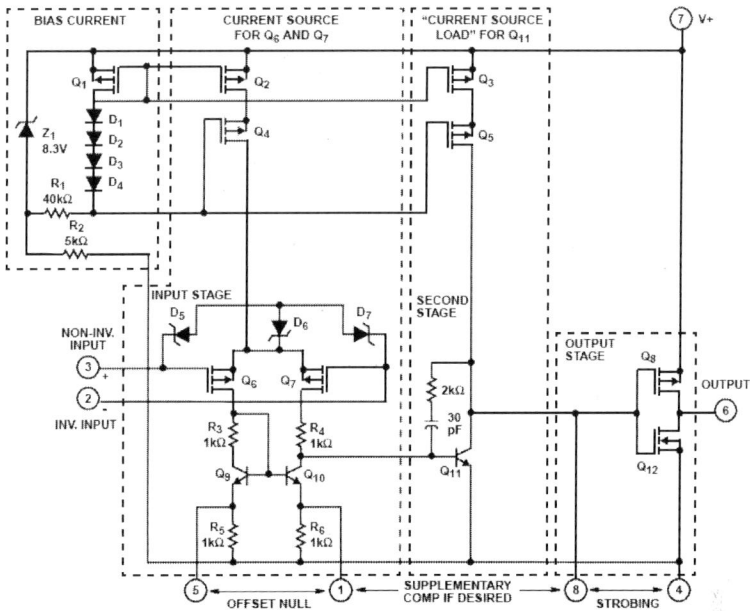

NOTE: Diodes D_5 Through D_7 Provide Gate Oxide Protection For MOSFET Input Stage.

Application Information

Circuit Description

Refer to the Block Diagram of the CA3160 series CMOS Operational Amplifiers. The input terminals may be operated down to 0.5V below the negative supply rail, and the output can be swung very close to either supply rail in many applications. Consequently, the CA3160 series circuits are ideal for single supply operation. Three class A amplifier stages, having the individual gain capability and current consumption shown in the Block Diagram provide the total gain of the CA3160. A biasing circuit provides two potentials for common use in the first and second stages. Terminals 8 and 1 can be used to supplement the internal phase compensation network if additional phase compensation or frequency roll-off is desired. Terminals 8 and 4 can also be used to strobe the output stage into a low quiescent current state. When Terminal 8 is tied to the negative supply rail (Terminal 4) by mechanical or electrical means, the output potential at Terminal 6 essentially rises to the positive supply-rail potential at Terminal 7. This condition of essentially zero current drain in the output stage under the strobed "OFF" condition can only be achieved when the ohmic load

resistance presented to the amplifier is very high (e.g., when the amplifier output is used to drive MOS digital circuits in comparator applications).

Input Stage - The circuit of the CA3160 is shown in the Schematic Diagram. It consists of a differential-input stage using PMOS field-effect transistors (Q_6, Q_7) working into a mirror-pair of bipolar transistors (Q_9, Q_{10}) functioning as load resistors together with resistors R_3 through R_6. The mirror-pair transistors also function as a differential-to-single-ended converter to provide base drive to the second-stage bipolar transistor (Q_{11}). Offset nulling, when desired, can be effected by connecting a 100,000Ω potentiometer across Terminals 1 and 5 and the potentiometer slider arm to Terminal 4. Cascode-connected PMOS transistors Q_2, Q_4, are the constant-current source for the input stage. The biasing circuit for the constant-current source is subsequently described. The small diodes D_5 through D_7 provide gate-oxide protection against high-voltage transients, including static electricity during handling for Q_6 and Q_7.

Second-Stage - Most of the voltage gain in the CA3160 is provided by the second amplifier stage, consisting of bipolar

SINGLE OPERATIONAL AMPLIFIERS

The LM741 series are general purpose operational amplifiers which feature improved performance over industry standards like the LM709. It is intended for a wide range of analog applications.
The high gain and wide range of operating voltage provide superior performance in integrator, summing amplifier, and general feedback applications.

FEATURES

- Short circuit protection
- Excellent temperature stability
- Internal frequency compensation
- High Input voltage range
- Null of offset

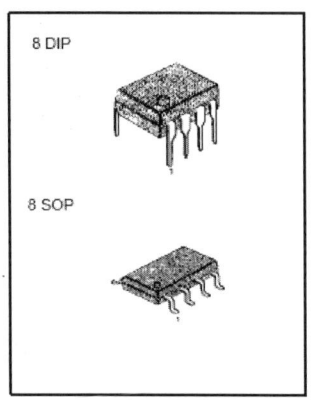

8 DIP

8 SOP

BLOCK DIAGRAM

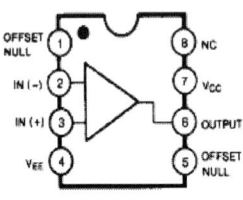

ORDERING INFORMATION

Device	Package	Operating Temperature
LM741N LM741EN	8 DIP	0 ~ + 70°C
LM741M LM741EM	8 SOP	0 ~ + 70°C
LM741IN LM741EIN	8 DIP	-40 ~ +85 °C
LM741IM LM741EIM	8 SOP	-40 ~ +85 °C

SCHEMATIC DIAGRAM

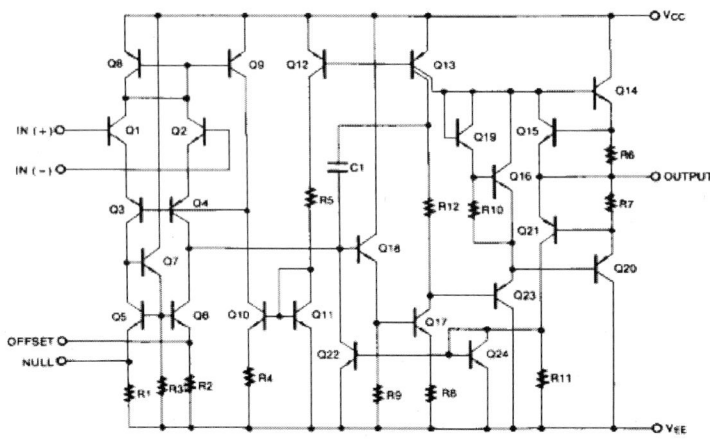

SINGLE OPERATIONAL AMPLIFIER

ABSOLUTE MAXIMUM RATINGS (T_A=25 °C)

Characteristic	Symbol	LM741	LM741E	LM741I	Unit
Supply Voltage	V_{CC}	±18	±22	±18	V
Differential Input Voltage	$V_{I(DIFF)}$	30	30	30	V
Input Voltage	V_I	±15	±15	±15	V
Output Short Circuit Duration		Indefinite	Indefinite	Indefinite	
Power Dissipation	P_D	500	500	500	mW
Operating Temperature Range	T_{OPR}	0 ~ +70	0 ~ +70	-40 ~ +85	°C
Storage Temperature Range	T_{STG}	-65 ~ +150	-65 ~ +150	-65 ~ +150	°C

ELECTRICAL CHARACTERISTICS
(V_{CC} = 15V, V_{EE} = -15V, T_A = 25 °C, unless otherwise specified)

Characteristic	Symbol	Test Conditions		LM741E			LM741/LM741I			Unit
				Min	Typ	Max	Min	Typ	Max	
Input Offset Voltage	V_{IO}	$R_g \leq 10K\Omega$						2.0	6.0	mV
		$R_g \leq 50\Omega$			0.8	3.0				
Input Offset Voltage Adjustment Range	$V_{IO(R)}$	V_{CC} = ±20V		±10				±15		mV
Input Offset Current	I_{IO}				3.0	30		20	200	nA
Input Bias Current	I_{BIAS}				30	80		80	500	nA
Input Resistance	R_I	V_{CC} = ±20V		1.0	6.0		0.3	2.0		MΩ
Input Voltage Range	$V_{I(R)}$			±12	±13		±12	±13		V
Large Signal Voltage Gain	G_V	$R_L \geq 2K\Omega$	V_{CC} = ±20V, $V_{O(P-P)}$ = ±15V	50						V/mV
			V_{CC} = ±15V, $V_{O(P-P)}$ = ±10V				20	200		
Output Short Circuit Current	I_{SC}			10	25	35		25		mA
Output Voltage Swing	$V_{O(P-P)}$	V_{CC} = ±20V	$R_L \geq 10K\Omega$	±16						V
			$R_L \geq 10K\Omega$	±15						
		V_{CC} = ±15V	$R_L \geq 10K\Omega$				±12	±14		
			$R_L \geq 10K\Omega$				±10	±13		
Common Mode Rejection Ratio	CMRR	$R_g \leq 10K\Omega$, V_{CM} = ±12V					70	90		dB
		$R_g \leq 50K\Omega$, V_{CM} = ±12V		80	95					
Power Supply Rejection Ratio	PSRR	V_{CC} = ±15V to V_{CC} = ±15V $R_g \leq 50\Omega$		86	96					dB
		V_{CC} = ±15V to V_{CC} = ±15V $R_g \leq 10K\Omega$					77	96		

ELECTRICAL CHARACTERISTICS (Continued)

Characteristic		Symbol	Test Conditions	LM741E			LM741/LM741I			Unit
				Min	Typ	Max	Min	Typ	Max	
Transient	Rise Time	t_R	Unity Gain		0.25	0.8		0.3		µs
Response	Overshoot	OS			6.0	20		10		%
Bandwidth		BW		0.43	1.5					MHz
Slew Rate		SR	Unity Gain	0.3	0.7			0.5		V/µs
Supply Current		I_{CC}	$R_L = \infty\Omega$					1.5	2.8	mA
Power Consumption		P_C	$V_{CC} = \pm20V$		80	150				mW
			$V_{CC} = \pm15V$					50	85	

ELECTRICAL CHARACTERISTICS

(-40 °C $\leq T_A \leq$ 85 °C for the KA741I °C $\leq T_A \leq$ 70 °C for the LM741 and LM741E. $V_{CC} = \pm15V$, unless otherwise specified)

Characteristic	Symbol	Test Conditions		LM741E			LM741/LM741I			Unit
				Min	Typ	Max	Min	Typ	Max	
Input Offset Voltage	V_{IO}	$R_S \leq 50\Omega$				4.0				mV
		$R_S \leq 10K\Omega$							7.5	
Input Offset Voltage Drift	$\Delta V_{IO}/\Delta T$				15					µV/°C
Input Offset Current	I_{IO}					70			300	nA
Input Offset Current Drift	$\Delta I_{IO}/\Delta T$					0.5				nA/°C
Input Bias Current	I_{BIAS}					0.21			0.8	µA
Input Resistance	R_I	$V_{CC} = \pm20V$		0.5						MΩ
Input Voltage Range	$V_{I(R)}$			±12	±13		±12	±13		V
Output Voltage Swing	$V_{O(P.P)}$	$V_{CC} = \pm20V$	$R_S \geq 10K\Omega$	±16						V
			$R_S \geq 2K\Omega$	±15						
		$V_{CC} = \pm15V$	$R_S \geq 10K\Omega$				±12	±14		
			$R_S \geq 2K\Omega$				±10	±13		
Output Short Circuit Current	I_{SC}			10		40	10		40	mA
Common Mode Rejection Ratio	CMRR	$R_S \leq 10K\Omega$, $V_{CM} = \pm12V$					70	90		dB
		$R_S \leq 50K\Omega$, $V_{CM} = \pm12V$		80	95					
Power Supply Rejection Ratio	PSRR	$V_{CC} = \pm20V$ to ±5V	$R_S \leq 50\Omega$	86	96					dB
			$R_S \leq 10K\Omega$				77	96		
Large Signal Voltage Gain	G_V	$R_S \geq 2K\Omega$	$V_{CC} = \pm20V$, $V_{O(P.P)} = \pm15V$	32						V/mV
			$V_{CC} = \pm15V$, $V_{O(P.P)} = \pm10V$				15			
			$V_{CC} = \pm15V$, $V_{O(P.P)} = \pm2V$	10						

MC14001B Series

B-Suffix Series CMOS Gates

MC14001B, MC14011B, MC14023B, MC14025B, MC14071B, MC14073B, MC14081B, MC14082B

The B Series logic gates are constructed with P and N channel enhancement mode devices in a single monolithic structure (Complementary MOS). Their primary use is where low power dissipation and/or high noise immunity is desired.

Features

- Supply Voltage Range = 3.0 Vdc to 18 Vdc
- All Outputs Buffered
- Capable of Driving Two Low–power TTL Loads or One Low–power Schottky TTL Load Over the Rated Temperature Range.
- Double Diode Protection on All Inputs Except: Triple Diode Protection on MC14011B and MC14081B
- Pin–for–Pin Replacements for Corresponding CD4000 Series B Suffix Devices
- Pb–Free Packages are Available*

MAXIMUM RATINGS (Voltages Referenced to V_{SS})

Symbol	Parameter	Value	Unit
V_{DD}	DC Supply Voltage Range	−0.5 to +18.0	V
V_{in}, V_{out}	Input or Output Voltage Range (DC or Transient)	−0.5 to V_{DD} + 0.5	V
I_{in}, I_{out}	Input or Output Current (DC or Transient) per Pin	± 10	mA
P_D	Power Dissipation, per Package (Note 1)	500	mW
T_A	Ambient Temperature Range	−55 to +125	°C
T_{stg}	Storage Temperature Range	−65 to +150	°C
T_L	Lead Temperature (8–Second Soldering)	260	°C

Maximum ratings are those values beyond which device damage can occur. Maximum ratings applied to the device are individual stress limit values (not normal operating conditions) and are not valid simultaneously. If these limits are exceeded, device functional operation is not implied, damage may occur and reliability may be affected.
1. Temperature Derating:
Plastic "P and D/DW" Packages: − 7.0 mW/°C From 65°C To 125°C

This device contains protection circuitry to guard against damage due to high static voltages or electric fields. However, precautions must be taken to avoid applications of any voltage higher than maximum rated voltages to this high–impedance circuit. For proper operation, V_{in} and V_{out} should be constrained to the range $V_{SS} \le (V_{in}$ or $V_{out}) \le V_{DD}$.

Unused inputs must always be tied to an appropriate logic voltage level (e.g., either V_{SS} or V_{DD}). Unused outputs must be left open.

*For additional information on our Pb–Free strategy and soldering details, please download the ON Semiconductor Soldering and Mounting Techniques Reference Manual, SOLDERRM/D.

ON Semiconductor®

http://onsemi.com

MARKING DIAGRAMS

 PDIP–14
P SUFFIX
CASE 646

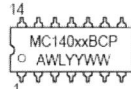
14
MC140xxBCP
AWLYYWW
1

 SOIC–14
D SUFFIX
CASE 751A

14
140xxB
AWLYWW
1

 TSSOP–14
DT SUFFIX
CASE 948G

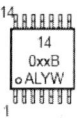
14
0xxB
ALYW
1

 SOEIAJ–14
F SUFFIX
CASE 965

14
MC140xxB
AWLYWW
1

xx = Specific Device Code
A = Assembly Location
WL, L = Wafer Lot
YY, Y = Year
WW, W = Work Week

DEVICE INFORMATION

Device	Description
MC14001B	Quad 2–Input NOR Gate
MC14011B	Quad 2–Input NAND Gate
MC14023B	Triple 3–Input NAND Gate
MC14025B	Triple 3–Input NOR Gate
MC14071B	Quad 2–Input OR Gate
MC14073B	Triple 3–Input AND Gate
MC14081B	Quad 2–Input AND Gate
MC14082B	Dual 4–Input AND Gate

ORDERING INFORMATION

See detailed ordering and shipping information in the package dimensions section on page 8 of this data sheet.

LOGIC DIAGRAMS

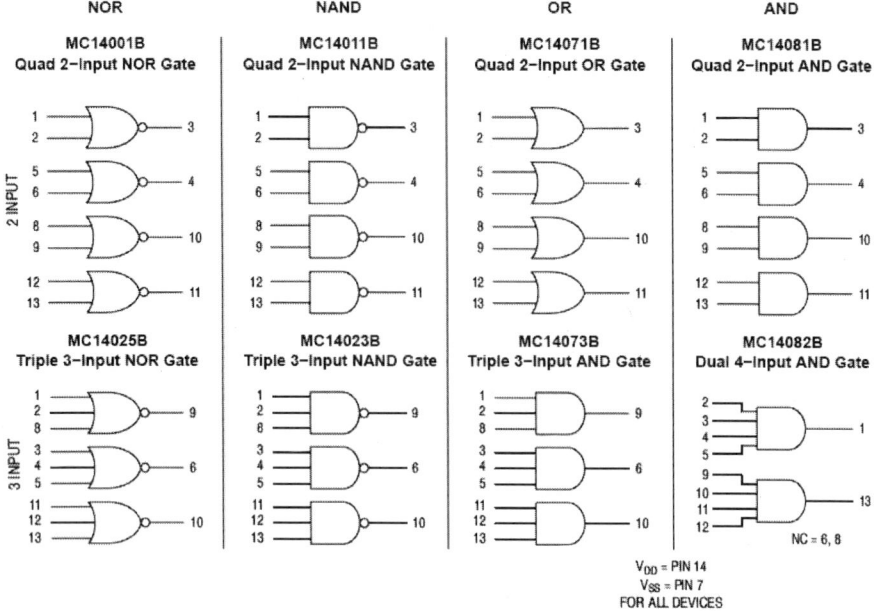

NOR	NAND	OR	AND
MC14001B Quad 2-Input NOR Gate	**MC14011B** Quad 2-Input NAND Gate	**MC14071B** Quad 2-Input OR Gate	**MC14081B** Quad 2-Input AND Gate
MC14025B Triple 3-Input NOR Gate	**MC14023B** Triple 3-Input NAND Gate	**MC14073B** Triple 3-Input AND Gate	**MC14082B** Dual 4-Input AND Gate

V_{DD} = PIN 14
V_{SS} = PIN 7
FOR ALL DEVICES

PIN ASSIGNMENTS

MC14001B
Quad 2-Input NOR Gate

IN 1_A	1	14	V_{DD}
IN 2_A	2	13	IN 2_D
OUT_A	3	12	IN 1_D
OUT_B	4	11	OUT_D
IN 1_B	5	10	OUT_C
IN 2_B	6	9	IN 2_C
V_{SS}	7	8	IN 1_C

MC14011B
Quad 2-Input NAND Gate

IN 1_A	1	14	V_{DD}
IN 2_A	2	13	IN 2_D
OUT_A	3	12	IN 1_D
OUT_B	4	11	OUT_D
IN 1_B	5	10	OUT_C
IN 2_B	6	9	IN 2_C
V_{SS}	7	8	IN 1_C

MC14023B
Triple 3-Input NAND Gate

IN 1_A	1	14	V_{DD}
IN 2_A	2	13	IN 3_C
IN 1_B	3	12	IN 2_C
IN 2_B	4	11	IN 1_C
IN 3_B	5	10	OUT_C
OUT_B	6	9	OUT_A
V_{SS}	7	8	IN 3_A

MC14025B
Triple 3-Input NOR Gate

IN 1_A	1	14	V_{DD}
IN 2_A	2	13	IN 3_C
IN 1_B	3	12	IN 2_C
IN 2_B	4	11	IN 1_C
IN 3_B	5	10	OUT_C
OUT_B	6	9	OUT_A
V_{SS}	7	8	IN 3_A

MC14071B
Quad 2-Input OR Gate

IN 1_A	1	14	V_{DD}
IN 2_A	2	13	IN 2_D
OUT_A	3	12	IN 1_D
OUT_B	4	11	OUT_D
IN 1_B	5	10	OUT_C
IN 2_B	6	9	IN 2_C
V_{SS}	7	8	IN 1_C

MC14073B
Triple 3-Input AND Gate

IN 1_A	1	14	V_{DD}
IN 2_A	2	13	IN 3_C
IN 1_B	3	12	IN 2_C
IN 2_B	4	11	IN 1_C
IN 3_B	5	10	OUT_C
OUT_B	6	9	OUT_A
V_{SS}	7	8	IN 3_A

MC14081B
Quad 2-Input AND Gate

IN 1_A	1	14	V_{DD}
IN 2_A	2	13	IN 2_D
OUT_A	3	12	IN 1_D
OUT_B	4	11	OUT_D
IN 1_B	5	10	OUT_C
IN 2_B	6	9	IN 2_C
V_{SS}	7	8	IN 1_C

MC14082B
Dual 4-Input AND Gate

OUT_A	1	14	V_{DD}
IN 1_A	2	13	OUT_B
IN 2_A	3	12	IN 4_B
IN 3_A	4	11	IN 3_B
IN 4_A	5	10	IN 2_B
NC	6	9	IN 1_B
V_{SS}	7	8	NC

NC = NO CONNECTION

ELECTRICAL CHARACTERISTICS (Voltages Referenced to V_{SS})

Characteristic		Symbol	V_{DD} Vdc	−55°C Min	−55°C Max	25°C Min	25°C Typ [2]	25°C Max	125°C Min	125°C Max	Unit
Output Voltage "0" Level $V_{in} = V_{DD}$ or 0		V_{OL}	5.0	−	0.05	−	0	0.05	−	0.05	Vdc
			10	−	0.05	−	0	0.05	−	0.05	
			15	−	0.05	−	0	0.05	−	0.05	
"1" Level $V_{in} = 0$ or V_{DD}		V_{OH}	5.0	4.95	−	4.95	5.0	−	4.95	−	Vdc
			10	9.95	−	9.95	10	−	9.95	−	
			15	14.95	−	14.95	15	−	14.95	−	
Input Voltage "0" Level (V_O = 4.5 or 0.5 Vdc) (V_O = 9.0 or 1.0 Vdc) (V_O = 13.5 or 1.5 Vdc)		V_{IL}	5.0	−	1.5	−	2.25	1.5	−	1.5	Vdc
			10	−	3.0	−	4.50	3.0	−	3.0	
			15	−	4.0	−	6.75	4.0	−	4.0	
"1" Level (V_O = 0.5 or 4.5 Vdc) (V_O = 1.0 or 9.0 Vdc) (V_O = 1.5 or 13.5 Vdc)		V_{IH}	5.0	3.5	−	3.5	2.75	−	3.5	−	Vdc
			10	7.0	−	7.0	5.50	−	7.0	−	
			15	11	−	11	8.25	−	11	−	
Output Drive Current (V_{OH} = 2.5 Vdc) Source (V_{OH} = 4.6 Vdc) (V_{OH} = 9.5 Vdc) (V_{OH} = 13.5 Vdc)		I_{OH}	5.0	− 3.0	−	− 2.4	− 4.2	−	− 1.7	−	mAdc
			5.0	− 0.64	−	− 0.51	− 0.88	−	− 0.36	−	
			10	− 1.6	−	− 1.3	− 2.25	−	− 0.9	−	
			15	− 4.2	−	− 3.4	− 8.8	−	− 2.4	−	
(V_{OL} = 0.4 Vdc) Sink (V_{OL} = 0.5 Vdc) (V_{OL} = 1.5 Vdc)		I_{OL}	5.0	0.64	−	0.51	0.88	−	0.36	−	mAdc
			10	1.6	−	1.3	2.25	−	0.9	−	
			15	4.2	−	3.4	8.8	−	2.4	−	
Input Current		I_{in}	15	−	± 0.1	−	±0.00001	± 0.1	−	± 1.0	μAdc
Input Capacitance ($V_{in} = 0$)		C_{in}	−	−	−	−	5.0	7.5	−	−	pF
Quiescent Current (Per Package)		I_{DD}	5.0	−	0.25	−	0.0005	0.25	−	7.5	μAdc
			10	−	0.5	−	0.0010	0.5	−	15	
			15	−	1.0	−	0.0015	1.0	−	30	
Total Supply Current [3] [4] (Dynamic plus Quiescent, Per Gate, C_L = 50 pF)		I_T	5.0	$I_T = (0.3\ \mu A/kHz)\ f + I_{DD}/N$							μAdc
			10	$I_T = (0.6\ \mu A/kHz)\ f + I_{DD}/N$							
			15	$I_T = (0.9\ \mu A/kHz)\ f + I_{DD}/N$							

2. Data labelled "Typ" is not to be used for design purposes but is intended as an indication of the IC's potential performance.
3. The formulas given are for the typical characteristics only at 25°C.
4. To calculate total supply current at loads other than 50 pF:

$$I_T(C_L) = I_T(50\ pF) + (C_L - 50)\ Vfk$$

where: I_T is in μA (per package), C_L in pF, $V = (V_{DD} - V_{SS})$ in volts, f in kHz is input frequency, and k = 0.001 x the number of exercised gates per package.

B-SERIES GATE SWITCHING TIMES

SWITCHING CHARACTERISTICS [5] (C_L = 50 pF, T_A = 25°C)

Characteristic	Symbol	V_{DD} Vdc	Min	Typ [6]	Max	Unit
Output Rise Time, All B–Series Gates	t_{TLH}					ns
$\quad t_{TLH}$ = (1.35 ns/pF) C_L + 33 ns		5.0	–	100	200	
$\quad t_{TLH}$ = (0.60 ns/pF) C_L + 20 ns		10	–	50	100	
$\quad t_{TLH}$ = (0.40 ns/PF) C_L + 20 ns		15	–	40	80	
Output Fall Time, All B–Series Gates	t_{THL}					ns
$\quad t_{THL}$ = (1.35 ns/pF) C_L + 33 ns		5.0	–	100	200	
$\quad t_{THL}$ = (0.60 ns/pF) C_L + 20 ns		10	–	50	100	
$\quad t_{THL}$ = (0.40 ns/pF) C_L + 20 ns		15	–	40	80	
Propagation Delay Time	t_{PLH}, t_{PHL}					ns
MC14001B, MC14011B only						
$\quad t_{PLH}, t_{PHL}$ = (0.90 ns/pF) C_L + 80 ns		5.0	–	125	250	
$\quad t_{PLH}, t_{PHL}$ = (0.36 ns/pF) C_L + 32 ns		10	–	50	100	
$\quad t_{PLH}, t_{PHL}$ = (0.26 ns/pF) C_L + 27 ns		15	–	40	80	
All Other 2, 3, and 4 Input Gates						
$\quad t_{PLH}, t_{PHL}$ = (0.90 ns/pF) C_L + 115 ns		5.0	–	160	300	
$\quad t_{PLH}, t_{PHL}$ = (0.36 ns/pF) C_L + 47 ns		10	–	65	130	
$\quad t_{PLH}, t_{PHL}$ = (0.26 ns/pF) C_L + 37 ns		15	–	50	100	
8–Input Gates (MC14068B, MC14078B)						
$\quad t_{PLH}, t_{PHL}$ = (0.90 ns/pF) C_L + 155 ns		5.0	–	200	350	
$\quad t_{PLH}, t_{PHL}$ = (0.36 ns/pF) C_L + 62 ns		10	–	80	150	
$\quad t_{PLH}, t_{PHL}$ = (0.26 ns/pF) C_L + 47 ns		15	–	60	110	

5. The formulas given are for the typical characteristics only at 25°C.
6. Data labelled "Typ" is not to be used for design purposes but is intended as an indication of the IC's potential performance.

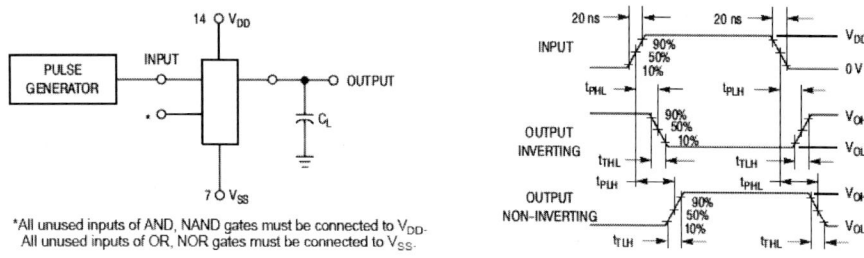

*All unused inputs of AND, NAND gates must be connected to V_{DD}.
All unused inputs of OR, NOR gates must be connected to V_{SS}.

Figure 1. Switching Time Test Circuit and Waveforms

CIRCUIT SCHEMATIC
NOR, OR GATES

MC14001B, MC14071B
One of Four Gates Shown

MC14025B
One of Three Gates Shown

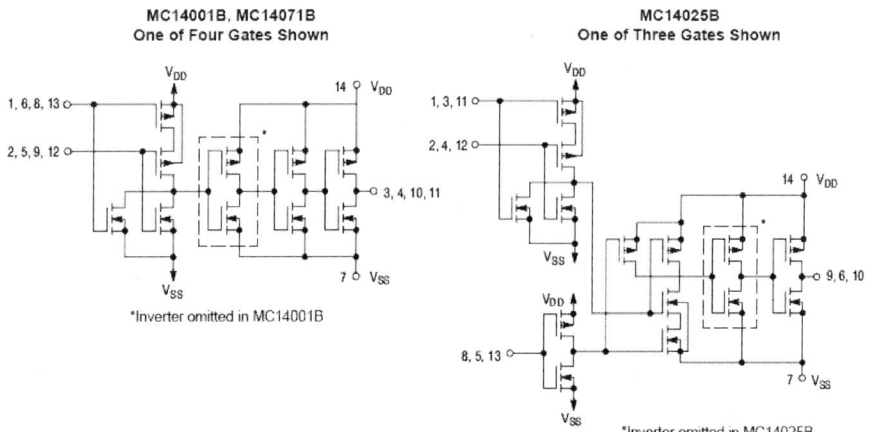

*Inverter omitted in MC14001B

*Inverter omitted in MC14025B

CIRCUIT SCHEMATIC
NAND, AND GATES

MC14023B, MC14073B
One of Three Gates Shown

MC14011B, MC14081B
One of Four Gates Shown

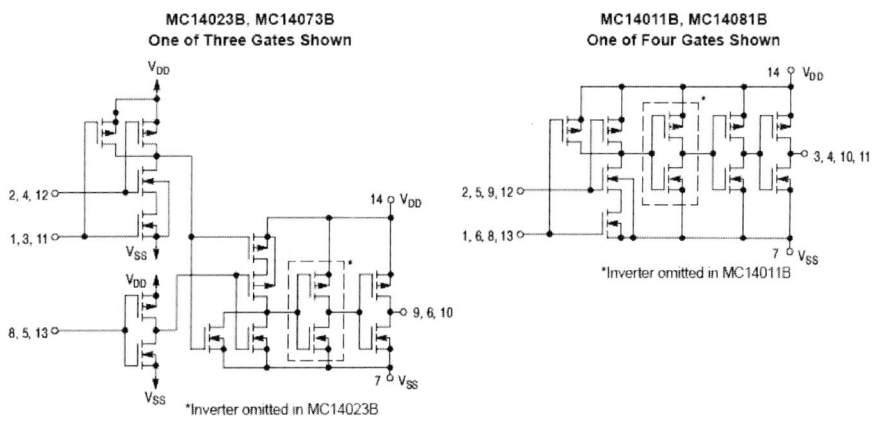

*Inverter omitted in MC14023B

*Inverter omitted in MC14011B

BCD Up/Down Counter

The MC14510B synchronous up/down BCD counter is constructed with MOS P–channel and N–channel enhancement mode devices in a monolithic structure. The counter consists of type D flip–flop stages with a gating structure to provide type T flip–flop capability.

This counter can be preset by applying the desired value in BCD to the Preset inputs (P1, P2, P3, P4) and then bringing the Preset Enable (PE) high. The direction of counting is controlled by applying a high (for up counting) or a low (for down counting) to the UP/DOWN input. The state of the counter changes on the positive transition of the clock input.

Cascading can be accomplished by connecting the Carry Out to the Carry In of the next stage while clocking each counter in parallel. The outputs (Q1, Q2, Q3, Q4) can be reset to a low state by applying a high to the Reset (R) pin.

This CMOS counter finds primary use in up/down and difference counting. Other applications include: (1) Frequency synthesizer applications where low power dissipation and/or high noise immunity is desired, (2) Analog–to–digital and digital–to–analog conversions, and (3) Magnitude and sign generation.

- Diode Protection on All Inputs
- Supply Voltage Range = 3.0 Vdc to 18 Vdc
- Internally Synchronous for High Speed
- Logic Edge–Clocked Design — Count Occurs on Positive Going Edge of Clock
- Asynchronous Preset Enable Operation
- Capable of Driving Two Low–power TTL Loads or One Low–power Schottky TTL Load Over the Rated Temperature Range.

MAXIMUM RATINGS* (Voltages Referenced to V_{SS})

Symbol	Parameter	Value	Unit
V_{DD}	DC Supply Voltage	– 0.5 to + 18.0	V
V_{in}, V_{out}	Input or Output Voltage (DC or Transient)	– 0.5 to V_{DD} + 0.5	V
I_{in}, I_{out}	Input or Output Current (DC or Transient), per Pin	± 10	mA
P_D	Power Dissipation, per Package†	500	mW
T_{stg}	Storage Temperature	– 65 to + 150	·C
T_L	Lead Temperature (8–Second Soldering)	260	·C

* Maximum Ratings are those values beyond which damage to the may occur.
† Temperature Derating:
  Plastic "P and D/DW" Packages: – 7.0 mW/·C From 65·C To 125·C
  Ceramic "L" Packages: – 12 mW/·C From 100·C To 125·C

TRUTH TABLE

Carry In	Up/Down	Preset Enable	Reset	Clock	Action
1	X	0	0	X	No Count
0	1	0	0	⌐	Count Up
0	0	0	0	⌐	Count Down
X	X	1	0	X	Preset
X	X	X	1	X	Reset

X = Don't Care
NOTE: When counting up, the Carry Out signal is normally high, and is low only when Q1 and Q4 are high and Carry In is low. When counting down, Carry Out is low only when Q1 through Q4 and Carry In are low.

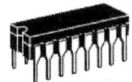

L SUFFIX
CERAMIC
CASE 620

P SUFFIX
PLASTIC
CASE 648

D SUFFIX
SOIC
CASE 751B

ORDERING INFORMATION

MC14XXXBCP	Plastic
MC14XXXBCL	Ceramic
MC14XXXBD	SOIC

T_A = – 55° to 125°C for all packages.

BLOCK DIAGRAM

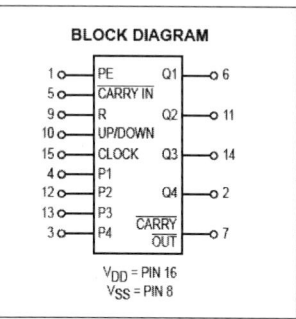

V_{DD} = PIN 16
V_{SS} = PIN 8

This device contains protection circuitry to guard against damage due to high static voltages or electric fields. However, precautions must be taken to avoid applications of any voltage higher than maximum rated voltages to this high–impedance circuit. For proper operation, V_{in} and V_{out} should be constrained to the range V_{SS} ≤ (V_{in} or V_{out}) ≤ V_{DD}.

Unused inputs must always be tied to an appropriate logic voltage level (e.g., either V_{SS} or V_{DD}). Unused outputs must be left open.

REV 3
1/94

© Motorola, Inc. 1995

MOTOROLA

ELECTRICAL CHARACTERISTICS (Voltages Referenced to V_{SS})

Characteristic	Symbol	V_{DD} Vdc	-55°C Min	-55°C Max	25°C Min	25°C Typ #	25°C Max	125°C Min	125°C Max	Unit
Output Voltage "0" Level $V_{in} = V_{DD}$ or 0	V_{OL}	5.0	—	0.05	—	0	0.05	—	0.05	Vdc
		10	—	0.05	—	0	0.05	—	0.05	
		15	—	0.05	—	0	0.05	—	0.05	
"1" Level $V_{in} = 0$ or V_{DD}	V_{OH}	5.0	4.95	—	4.95	5.0	—	4.95	—	Vdc
		10	9.95	—	9.95	10	—	9.95	—	
		15	14.95	—	14.95	15	—	14.95	—	
Input Voltage "0" Level ($V_O = 4.5$ or 0.5 Vdc) ($V_O = 9.0$ or 1.0 Vdc) ($V_O = 13.5$ or 1.5 Vdc)	V_{IL}	5.0	—	1.5	—	2.25	1.5	—	1.5	Vdc
		10	—	3.0	—	4.50	3.0	—	3.0	
		15	—	4.0	—	6.75	4.0	—	4.0	
"1" Level ($V_O = 0.5$ or 4.5 Vdc) ($V_O = 1.0$ or 9.0 Vdc) ($V_O = 1.5$ or 13.5 Vdc)	V_{IH}	5.0	3.5	—	3.5	2.75	—	3.5	—	Vdc
		10	7.0	—	7.0	5.50	—	7.0	—	
		15	11	—	11	8.25	—	11	—	
Output Drive Current Source ($V_{OH} = 2.5$ Vdc) ($V_{OH} = 4.6$ Vdc) ($V_{OH} = 9.5$ Vdc) ($V_{OH} = 13.5$ Vdc)	I_{OH}	5.0	-3.0	—	-2.4	-4.2	—	-1.7	—	mAdc
		5.0	-0.64	—	-0.51	-0.88	—	-0.36	—	
		10	-1.6	—	-1.3	-2.25	—	-0.9	—	
		15	-4.2	—	-3.4	-8.8	—	-2.4	—	
Sink ($V_{OL} = 0.4$ Vdc) ($V_{OL} = 0.5$ Vdc) ($V_{OL} = 1.5$ Vdc)	I_{OL}	5.0	0.64	—	0.51	0.88	—	0.36	—	mAdc
		10	1.6	—	1.3	2.25	—	0.9	—	
		15	4.2	—	3.4	8.8	—	2.4	—	
Input Current	I_{in}	15	—	±0.1	—	±0.00001	±0.1	—	±1.0	μAdc
Input Capacitance ($V_{in} = 0$)	C_{in}	—	—	—	—	5.0	7.5	—	—	pF
Quiescent Current (Per Package)	I_{DD}	5.0	—	5.0	—	0.005	5.0	—	150	μAdc
		10	—	10	—	0.010	10	—	300	
		15	—	20	—	0.015	20	—	600	
Total Supply Current**† (Dynamic plus Quiescent, Per Package) ($C_L = 50$ pF on all outputs, all buffers switching)	I_T	5.0 10 15	$I_T = (0.58 \text{ μA/kHz}) f + I_{DD}$ $I_T = (1.20 \text{ μA/kHz}) f + I_{DD}$ $I_T = (1.70 \text{ μA/kHz}) f + I_{DD}$							μAdc

#Data labelled "Typ" is not to be used for design purposes but is intended as an indication of the IC's potential performance.

** The formulas given are for the typical characteristics only at 25°C.

† To calculate total supply current at loads other than 50 pF:

$$I_T(C_L) = I_T(50 \text{ pF}) + (C_L - 50) \text{ Vfk}$$

where: I_T is in μA (per package), C_L in pF, $V = (V_{DD} - V_{SS})$ in volts, f in kHz is input frequency, and $k = 0.001$.

PIN ASSIGNMENT

PE	1 ●	16	V_{DD}
Q4	2	15	C
P4	3	14	Q3
P1	4	13	P3
CARRY IN	5	12	P2
Q1	6	11	Q2
CARRY OUT	7	10	U/D
V_{SS}	8	9	R

SWITCHING CHARACTERISTICS (C_L = 50 pF, T_A = 25°C, See Figure 2)

Characteristic	Symbol	V_{DD}	All Types Min	All Types Typ #	All Types Max	Unit
Output Rise and Fall Time t_{TLH}, t_{THL} = (1.5 ns/pF) C_L + 25 ns t_{TLH}, t_{THL} = (0.75 ns/pF) C_L + 12.5 ns t_{TLH}, t_{THL} = (0.55 ns/pF) C_L + 9.5 ns	t_{TLH}, t_{THL}	5.0 10 15	— — —	100 50 40	200 100 80	ns
Propagation Delay Time Clock to Q t_{PLH}, t_{PHL} = (1.7 ns/pF) C_L + 230 ns t_{PLH}, t_{PHL} = (0.66 ns/pF) C_L + 97 ns t_{PLH}, t_{PHL} = (0.5 ns/pF) C_L + 75 ns	t_{PLH}, t_{PHL}	5.0 10 15	— — —	315 130 100	630 260 200	ns
Clock to Carry Out t_{PLH}, t_{PHL} = (1.7 ns/pF) C_L + 230 ns t_{PLH}, t_{PHL} = (0.66 ns/pF) C_L + 97 ns t_{PLH}, t_{PHL} = (0.5 ns/pF) C_L + 75 ns	t_{PLH}, t_{PHL}	5.0 10 15	— — —	315 130 100	630 260 200	ns
Carry In to Carry Out t_{PLH}, t_{PHL} = (1.7 ns/pF) C_L + 230 ns t_{PLH}, t_{PHL} = (0.66 ns/pF) C_L + 47 ns t_{PLH}, t_{PHL} = (0.5 ns/pF) C_L + 35 ns	t_{PLH}, t_{PHL}	5.0 10 15	— — —	180 80 60	360 160 120	ns
Preset or Reset to Q t_{PLH}, t_{PHL} = (1.7 ns/pF) C_L + 230 ns t_{PLH}, t_{PHL} = (0.66 ns/pF) C_L + 97 ns t_{PLH}, t_{PHL} = (0.5 ns/pF) C_L + 75 ns	t_{PLH}, t_{PHL}	5.0 10 15	— — —	315 130 100	630 260 200	ns
Preset or Reset to Carry Out t_{PLH}, t_{PHL} = (1.7 ns/pF) C_L + 465 ns t_{PLH}, t_{PHL} = (0.66 ns/pF) C_L + 192 ns t_{PLH}, t_{PHL} = (0.5 ns/pF) C_L + 125 ns	t_{PLH}, t_{PHL}	5.0 10 15	— — —	550 225 150	1100 450 300	ns
Reset Pulse Width	$t_{w(H)}$	5.0 10 15	360 210 160	180 105 80	— — —	ns
Clock Pulse Width	$t_{w(H)}$	5.0 10 15	350 170 140	200 100 75	— — —	ns
Clock Pulse Frequency	f_{cl}	5.0 10 15	— — —	3.0 6.0 8.0	1.5 3.0 4.0	MHz
Preset or Reset Removal Time The Preset or Reset Signal must be low prior to a positive–going transition of the clock.	t_{rem}	5.0 10 15	650 230 180	325 115 90	— — —	ns
Clock Rise and Fall Time	t_{TLH}, t_{THL}	5.0 10 15	— — —	— — —	15 5 4	µs
Setup Time Carry In to Clock	t_{su}	5.0 10 15	260 120 100	130 60 50	— — —	ns
Hold Time Clock to Carry In	t_h	5.0 10 15	0 10 10	– 50 – 15 – 5	— — —	ns
Setup Time Up/Down to Clock	t_{su}	5.0 10 15	500 200 175	250 100 75	— — —	ns
Hold Time Clock to Up/Down	t_h	5.0 10 15	– 70 – 30 – 20	– 140 – 80 – 50	— — —	ns
Setup Time Pn to PE	t_{su}	5.0 10 15	– 50 – 30 – 25	– 100 – 65 – 55	— — —	ns
Hold Time PE to Pn	t_h	5.0 10 15	480 410 410	240 205 205	— — —	ns
Preset Enable Pulse Width	t_{WH}	5.0 10 15	200 100 80	100 50 40	— — —	ns

* The formulas given are for the typical characteristics only at 25°C.
#Data labelled "Typ" is not to be used for design purposes but is intended as an indication of the IC's potential performance.

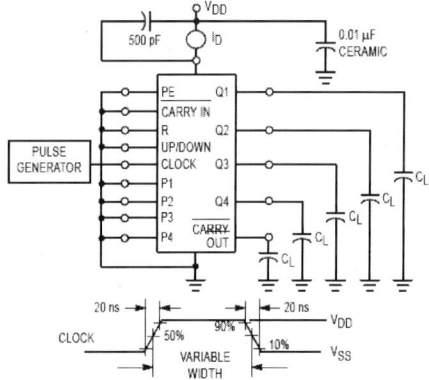

Figure 1. Power Dissipation Test Circuit and Waveform

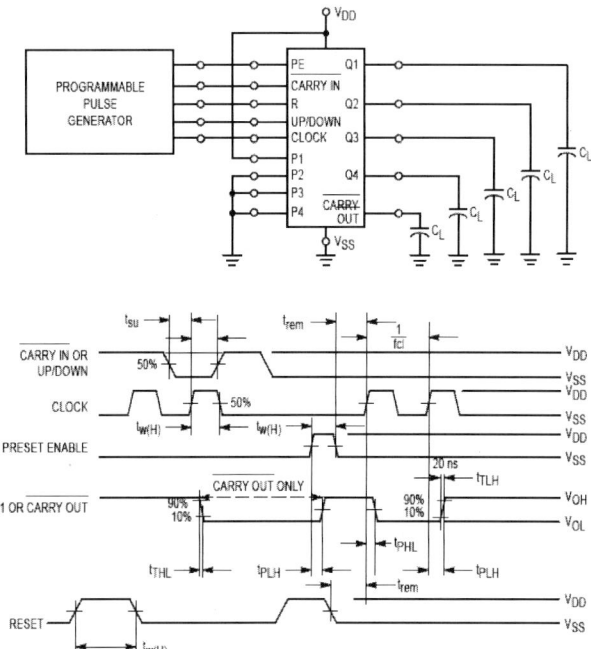

Figure 2. Switching Time Test Circuit and Waveforms

LOGIC DIAGRAM

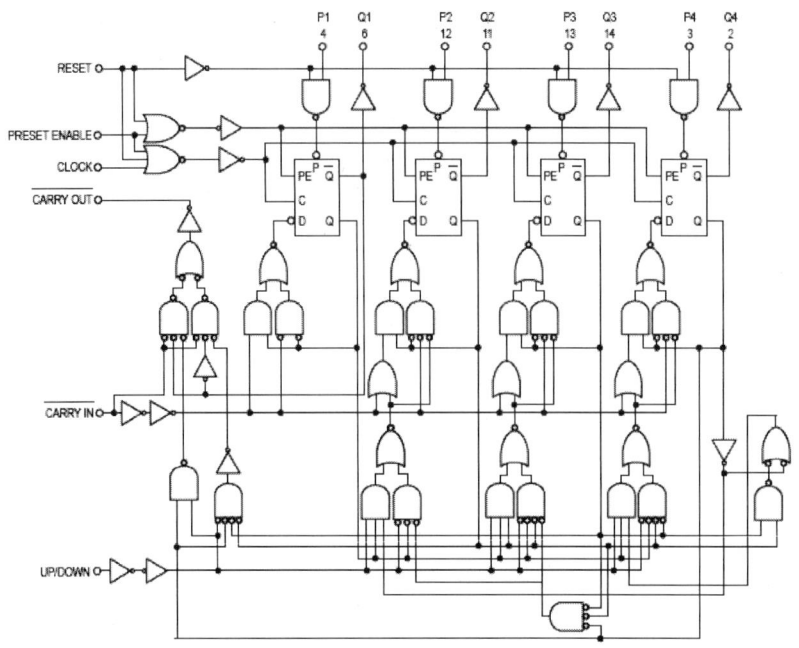

STATE DIAGRAM FOR UP COUNTING

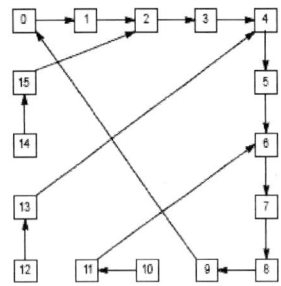

STATE DIAGRAM FOR DOWN COUNTING

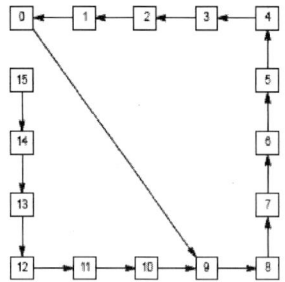

PIN DESCRIPTIONS

INPUTS

P1, P2, P3, P4, Preset Inputs (Pins 4, 12, 13, 3) — Data on these inputs is loaded into the counter when PE is taken high.

Carry In, (Pin 5) — Active–low input used when cascading stages. Usually connected to Carry Out of the previous stage. While high, clock is inhibited.

Clock, (Pin 15) — BCD data is incremented or decremented, depending on the direction of count, on the positive transition of this signal.

OUTPUTS

Q1, Q2, Q3, Q4, BCD outputs (Pins 6, 11, 14, 2) — BCD data is present on these outputs with Q1 corresponding to the least significant bit.

Carry Out, (Pin 7) — Used when cascading stages, this pin is usually connected to Carry In of the next stage. This synchronous output is active low and may also be used to indicate terminal count.

CONTROLS

PE, Preset Enable (Pin 1) — Asynchronously loads data on the Preset Inputs. This pin is active high and will inhibit the clock when high.

R, Reset, (Pin 9) — Asynchronously resets the Q outputs to a low state. This pin is active high and will inhibit the clock when high.

Up/Down, (Pin 10) — Controls the direction of count: high for up count, low for down count.

SUPPLY PINS

V$_{SS}$, Negative Supply Voltage, (Pin 8) — This pin is usually connected to ground.

V$_{DD}$, Positive Supply Voltage, (Pin 16) — This pin is connected to a positive supply voltage ranging from 3.0 Vdc to 18.0 Vdc.

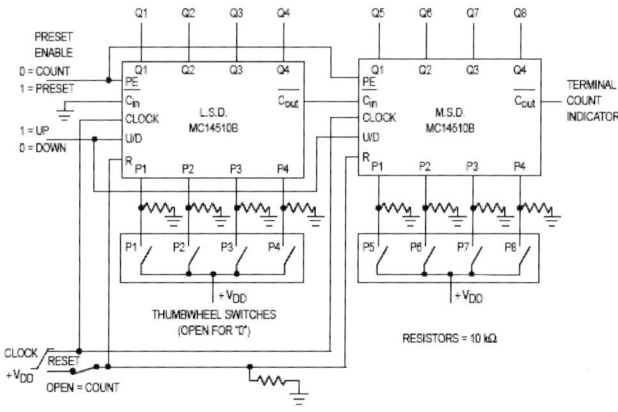

Note: The Least Significant Digit (L.S.D.) counts from a preset value once Preset Enable (PE) goes low. The Most Significant Digit (M.S.D.) does not change while C$_{in}$ is high. When the count of the L.S.D. reaches 0 (count down mode) or reaches 9 (count up mode), C$_{out}$ goes low for one complete clock cycle, thus allowing the next counter to decrement/increment one count. The L.S.D. now counts through another cycle (10 clock pulses) and the above cycle is repeated.

Figure 3. Presettable Cascaded 8–Bit Up/Down Counter

TIMING DIAGRAM FOR THE PRESETTABLE
CASCADED 8-BIT UP/DOWN COUNTER

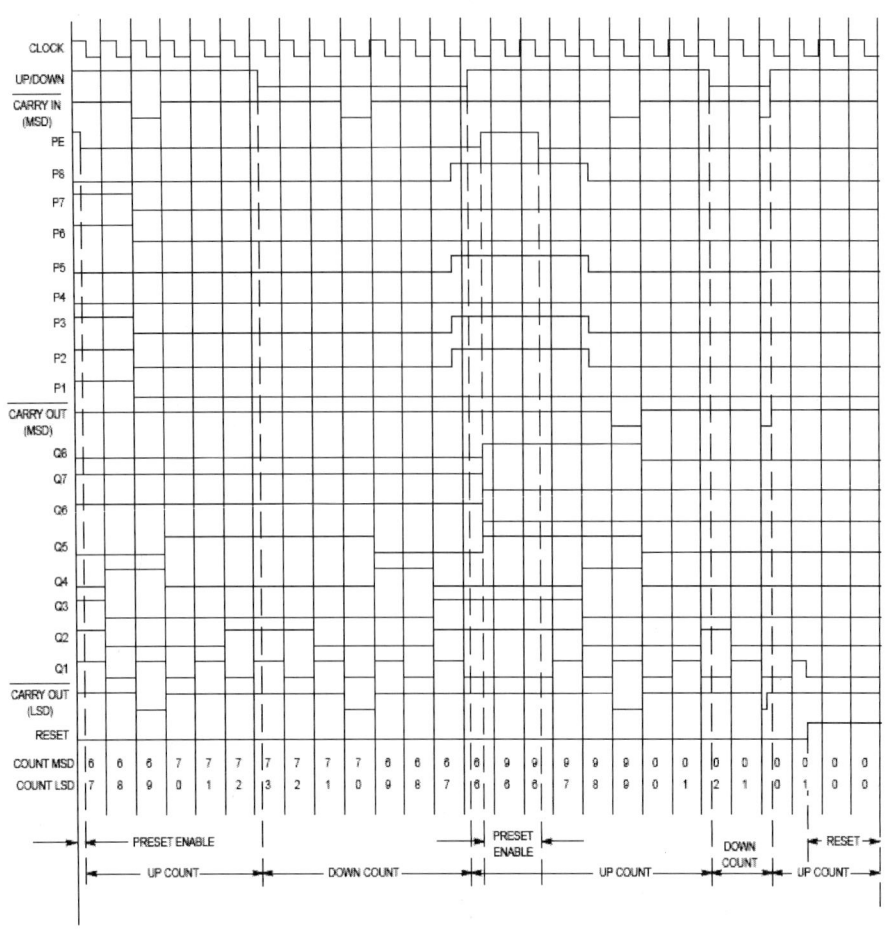

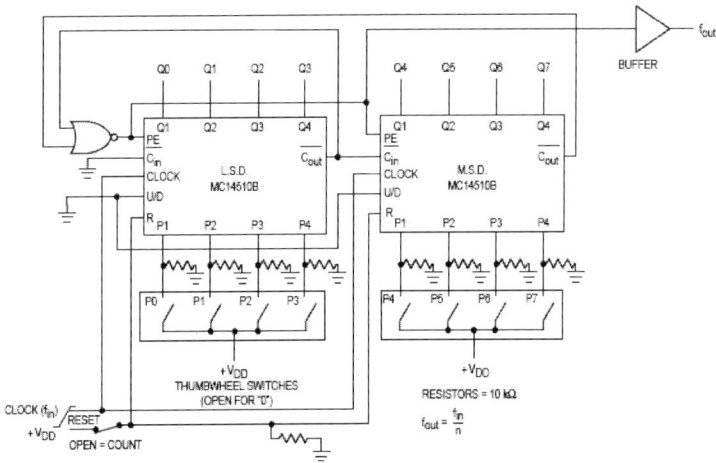

Note: The programmable frequency divider can be set by applying the desired divide ratio, in BCD, to the preset inputs. For example, the maximum divide ratio of 99 may be obtained by applying a 10011001 to the preset inputs P0 to P7. For this divide operation, both counters should be configured in the count down mode. The divide ratio of zero is an undefined state and should be avoided.

Figure 4. Programmable Cascaded Frequency Divider

Red GaAsP 0.5-Inch 7-Segment Numeric LED Displays

Optoelectronic Products

FND500, FND507 FND560, FND567

General Description

The FND500, FND507, FND560 and FND567 are red GaAsP single-digit 7-Segment LED displays with a 0.5-inch character height. These displays are designed for applications in which the viewer is within twenty feet of the display.

Low Forward Voltage — Typically V_F = 1.7 V
Fits Standard DIP Sockets with 0.6-Inch Pin Row
Maximized Contrast Ratio With Integral Lens Cap
Horizontal Stacking 0.6-Inch Minimum,
    1-Inch Typical
FND560/567 Suitable For Use In High
    Ambient Light
FND500 Common Cathode, Right-Hand
    Decimal Point
FND507 Common Anode, Right-Hand Decimal Point
FND560 Common Cathode, Right-Hand Decimal
    Point, High Brightness
FND567 Common Anode, Right-Hand Decimal
    Point, High Brightness

Absolute Maximum Ratings

Maximum Temperature and Humidity

Storage Temperature	$-25°C$ to $+85°C$
Operating Temperature	$-25°C$ to $+85°C$
Pin Temperature (Soldering, 5 s)	$260°C$
Relative Humidity at 65°C	98%

Maximum Voltage and Currents

V_R	Reverse Voltage	3.0 V
I_F	Average Forward dc Current / Segment or Decimal Point	25 mA
	Derate from 25°C Ambient Temperature	0.3 mA / °C
I_{pk}	Peak Forward Current Segment or Decimal Point (100 μs pulse width) 1000 pps, T_A = 25°C	200 mA

Package Outline

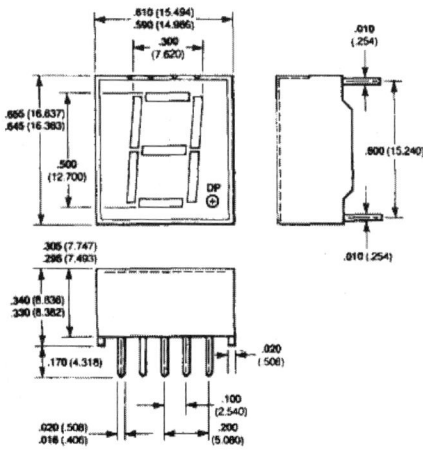

Notes
All dimensions in inches bold and millimeters (parentheses)
Tolerance unless specified = ±.015 (±.381)

Connection Diagram
Typical Electrical
Characteristics

FND500, FND507
FND560, FND567

Pin Connections
(Front View)

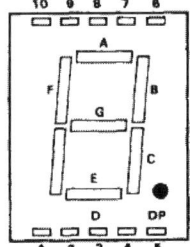

Pin	FND507/567	FND500/560
1	Segment E	Segment E
2	Segment D	Segment D
3	Common Anode	Common Cathode
4	Segment C	Segment C
5	Decimal Point	Decimal Point
6	Segment B	Segment B
7	Segment A	Segment A
8	Common Anode	Common Cathode
9	Segment F	Segment F
10	Segment G	Segment G

Electrical and Radiant Characteristics $T_A = 25°C$

Symbol	Characteristic	Min	Typ	Max	Units	Test Conditions
V_F	Forward Voltage	1.5	1.7	2.0	V	$I_F = 20$ mA
BV_R	Reverse Breakdown Voltage	3.0	12		V	$I_R = 1.0$ mA
I_O	Axial Luminous Intensity, Average					
	Each Segment					
	FND500, FND507	300	600		μcd	$I_F = 20$ mA
	FND560, FND567	740	1200		μcd	$I_F = 20$ mA
ΔI_O	Intensity Matching, Segment-to-Segment		±33		%	$I_F = 20$ mA
	Intensity Matching Within One Intensity Class		±20		%	$I_F = 20$ mA, all segments at once
L_O	Average Segment Luminance					
	FND500, FND507		35		ftL	$I_F = 20$ mA
	FND560, FND567		70		ftL	$I_F = 20$ mA
$\theta_{1/2}$	Viewing Angle to Half Intensity		±27		degrees	
λ_{pk}	Peak Wavelength		665		nm	$I_F = 20$ mA

부록 2. TTL and CMOS Logic listings

Type	Description	Pins	74 HC	74 HCT	74 AC	74 ACT	74 F	4000	74 ALS	74 AS	74	74 LS
00	Quad 2 Input NAND Gate	14	X	X	X	X	X		X	X	X	X
01	Quad 2 Input NAND Gate (OC)	14										
02	Quad 2 Input Positive NOR Gate	14	X		X	X	X		X		X	X
03	Quad 2 Input Positive NAND Gate	14							X			X
04	Hex Inverter	14	X	X	X	X	X		X	X	X	X
05	Hex Inverter (OC)	14		X	X				X		X	X
06	Hex Inverter Buffer/Driver (OC)	14									X	X
07	30V, 40mA Hex Buffer Driver (OC)	14									X	X
08	Quad 2 Input Positive AND Gate	14	X	X	X	X	X		X	X		X
09	Quad 2 Input Positive AND Gate (OC)	14							X			X
10	Triple 3 Input Positive NAND Gate	14			X	X	X		X	X	X	X
11	Triple 3 Input Positive AND Gate	14			X		X		X			X
12	Triple 3−Input Nand (Open Collector)	14										
13	Dual 4−Input Schmitt Trig. Nand	14										
14	Hex Schmitt Trigger	14	X	X	X	X	X		X		X	X
16	Hex Inverter Buffer/Driver (OC)	14								X		
17	15V, 40mA Hex Buffer/Driver (OC)	14								X		
19	Hex Schmitt−Trigger Inverter	14										X
20	Dual 4 Input Positive NAND Gate	14			X		X		X			X
21	Dual 4 Input Positive AND Gate	14							X	X		X
22	Dual 4−Input Nand Gate (Open Collector)	14										
23	Expandable Dual 4−Input Nor Gate	16										
25	Dual 4 Input Positive NOR Gate	14								X		
26	Quad 2 Input NAND Gate (OC)	14										X
27	Triple 3 Input NOR Gate	14							X	X		X
28	Quad 2 Input NOR Buffer	14										
30	8 Input Positive NAND Gate	14							X	X		X
31	Delay Element	16										X
32	Quad 2 Input Positive OR Gate	14	X	X	X	X	X		X	X	X	X

Type	Description	Pins	74 HC	74 HCT	74 AC	74 ACT	74 F	4000	74 ALS	74 AS	74	74 LS
33	Quad 2 Input NOR Buffer (OC)	14							X			X
34	Hex Non-Inverter	14								X		
35	Hex Non-Inverter (OC)	14							X			
37	Quad 2 Input Positive NAND Buffer	14							X			X
38	Quad 2 Input NAND Buffer (OC)	14				X			X	X		X
40	Dual 4-Input Nand Buffer Gates	14										
41	Bcd-Todecimal Decoder/Driver (Nixie)	16									X	
42	Excess-3 Decimal Decoder	16										
43	Excess-3-Gray Decimal Decoder	16										
44	Excess-3-Gray Decimal Decoder	16										
45	BCD to Decimal Decoder/Driver (OC)	16									X	
46	BCD to 7 Segment Decoder/Driver (30V)	16										
47	BCD to 7 Segment Decoder/Driver (15V)	16									X	X
48	BCD to 7 Segment Decoder/Driver	16										
50	Xpan. Dual 2-Wide 2in A-O-I Gate	14										
51	Dual 2 Wide 2 Input AND/OR Inv.Gate	14										X
53	Xpan. 4-Wide 2in And/Or Inverter	14										
54	4 Wide 2-Input And/Or Inverter	14										
59	2 Wide 2-3 In And/Or Inverter Gate	14										
60	Dual 4-Input Expanders	14										
70	Edge Triggered Jk Flip Flop	14										
72	J-K Master Slave Flip Flop	14										
73	Dual J-K Master Slave Flip/Flop	14										X
74	Dual D Type Edge Triggered Flip/Flop	14	X	X	X	X	X		X	X		X
75	Quad Bi-Stable Latch	16										X
76	Dual J-K Master Slave Flip/Flop	16										
77	4 Bit Bi-Stable Latch	14										
79	Dual D Flip Flop	14										
80	Gated Full Adder	14										
82	2-Bit Full Adder	14										
83	4 Bit Binary Full Adder	16										

Type	Description	Pins	74 HC	74 HCT	74 AC	74 ACT	74 F	4000	74 ALS	74 AS	74	74 LS
85	4 Bit Magnitude Comparator	16										X
86	Quad 2 Input XOR Gate	14	X		X		X		X			X
89	64−Bit RAM	16										
90	Decade Counter	14										X
91	8−Bit Shift Register	14										
92	Divide−by−12 Decade Counter	14										X
93	4 Bit Binary Counter	14										X
94	4−Bit Shift Reg. Par. In/Ser. Out	16										
95	4−Bit Right Shift Left Shift Reg.	14										
96	5−Bit Par. In Par. Outshift Reg.	16										
97	6 Bit Asynchronous Binary Rate Multiplier	16								X		
100	Dual 4−Bit Bistable Latch	24										
104	Gated J−K Master Slave Flip Flop	14										
107	Dual J−K Master Slave Flip/Flop	14								X		X
109	Dual J−K Positive Edge Triggered Flip/Flop	16			X	X	X		X			X
112	Dual J−K Negative Edge Triggered Flip/Flop	16					X		X			X
121	One Shot Multivibrator	14								X		
122	Retriggerable Monostable Multivibrator	14										X
123	Dual One Shot Multivibrator	16	X							X		X
124	Dual Voltage Controlled Oscillator	16										
125	Tri−State Quad Buffer	14	X		X	X	X		X			X
126	Tri−State Quad Buffer	14	X									X
128	Quad 2 Input NOR Line Driver	14								X		
132	Quad Schmitt Trigger	14	X			X						X
133	13 Input NAND Gate	16							X			
136	Quad Exclusive OR Gate (OC)	14										X
137	3/8 Decoder Multiplexer	16							X			
138	Expandable 3/8 Decoder	16	X	X	X	X	X		X			X
139	Dual 2 to 4 Decoder Demultiplexer	16			X	X	X		X			X

Type	Description	Pins	74 HC	74 HCT	74 AC	74 ACT	74 F	4000	74 ALS	74 AS	74	74 LS
140	Dual 4 Input NAND Line Driver	14										
141	Bcd-Todecimal Decoder/Driver (Nixie)	16									X	
145	BCD to Decimal Decoder/Driver (OC)	16									X	X
147	10 to 4 Line Priority Encoder	16										
148	8 to 3 Priority Encoder	16					X					X
150	16 Line Multiplexer	24									X	
151	8 Line Multiplexer	16	X		X	X	X		X			X
152	8-Channel Data Selector	14										
153	Dual 4 Input Multiplexer	16			X	X	X		X			X
154	4 Line to 16 Line Decoder Demultiplexer	24	X								X	
155	Dual 2 to 4 Demultiplexer	18										X
156	Dual 2 to 4 Demultiplexer (OC)	16							X			X
157	Quad 2 Input Multiplexer (9322)	16	X		X	X	X		X	X		X
158	Quad 2 Input Data Selector/Multiplexer	16				X			X	X		X
159	4 to 16 Line Decoder	24									X	
160	Synchronous 4 Bit Decade Counter	16										
121	One Shot Multivibrator	14									X	
122	Retriggerable Monostable Multivibrator	14										X
123	Dual One Shot Multivibrator	16	X								X	X
124	Dual Voltage Controlled Oscillator	16										
125	Tri-State Quad Buffer	14	X		X	X	X		X			X
126	Tri-State Quad Buffer	14	X									X
128	Quad 2 Input NOR Line Driver	14									X	
132	Quad Schmitt Trigger	14	X				X					X
133	13 Input NAND Gate	16							X			
136	Quad Exclusive OR Gate (OC)	14										X
137	3/8 Decoder Multiplexer	16							X			
138	Expandable 3/8 Decoder	16	X	X	X	X	X		X			X
139	Dual 2 to 4 Decoder Demultiplexer	16			X	X	X		X			X
140	Dual 4 Input NAND Line Driver	14										

Type	Description	Pins	74 HC	74 HCT	74 AC	74 ACT	74 F	4000	74 ALS	74 AS	74	74 LS
141	Bcd—Todecimal Decoder/Driver (Nixie)	16								X		
145	BCD to Decimal Decoder/Driver (OC)	16								X		X
147	10 to 4 Line Priority Encoder	16										
148	8 to 3 Priority Encoder	16					X					X
150	16 Line Multiplexer	24							X			
151	8 Line Multiplexer	16	X		X	X	X		X			X
152	8—Channel Data Selector	14										
153	Dual 4 Input Multiplexer	16			X	X	X		X			X
154	4 Line to 16 Line Decoder Demultiplxer	24	X						X			
155	Dual 2 to 4 Demultiplexer	18										X
156	Dual 2 to 4 Demultiplexer (OC)	16							X			X
157	Quad 2 Input Multiplexer (9322)	16	X		X	X	X		X	X		X
158	Quad 2 Input Data Selector/Multiplexer	16				X			X	X		X
159	4 to 16 Line Decoder	24								X		
160	Synchronous 4 Bit Decade Counter	16										
241	Octal Tri—State Buffer	20			X	X			X			X
243	Quad Bus Transceiver	14							X			X
244	Octal Tri—State Buffer	20	X	X	X	X	X		X	X		X
245	Octal Tri—State Transceiver	20	X	X	X	X	X		X	X		X
247	BCD to 7—segment Decoder/Driver	16										X
251	Data Selector/Multiplexer	16	X		X	X	X		X			X
253	Tri—State Dual 4 Input Multiplexer	16			X	X	X		X			X
257	Tri—State Quad 2 Input Multiplexer	16			X	X	X		X	X		X
258	Tri—State Quad 2 Input Multiplexer	16				X			X	X		X
259	8 Bit Addressable Latch	16	X						X			X
260	Dual 5 Bit Input NOR Gate	14										
265	Quad Complementary—Output Elements	16										
266	Quad 2—Input exclusive—NOR Gate, O/C outputs	14										X
269	8 Bit Up/Down Counter	24					X					

Type	Description	Pins	74 HC	74 HCT	74 AC	74 ACT	74 F	4000	74 ALS	74 AS	74	74 LS
273	Octal D Type Flip/Flop with Clear	20	X	X	X	X	X		X			X
276	Quad J−K Flip Flop	20										
279	Quad R−S Latch	16										X
280	9 Bit Parity Generator/Checker	14			X		X		X	X		X
283	4 Bit Binary Adder	16										X
284	Tri−State 4−Bit Multiplexer (Multiplier?)	16										
285	Tri−State 4−Bit Multiplexer (Multiplier?)	16										
286	9 Bit Parity Generator/Checker	14								X		
292	16 Bit Programmable Frequency Divider	16										X
293	4−bit Binary Counters	14										X
294	16 Bit Programmable Frequency Divider	16										X
297	Digital Phase Locked Loop	16										X
298	Quad 2−input Multiplexers with storage	16										X
299	8 Bit Shift/Storage Register	20			X	X			X			X
321	Crystal Oscillator	16										
323	8 Bit Shift/Storage Register	20				X			X			
348	8 to 3 Priority Encoder	16										X
351	Dual Data Selector/Multiplexer T.S. Output	20										
365	Tri−State Hex Buffer	16										X
366	Hex Inverting Bus Drivers	16										
367	Tri−State Hex Buffer	16										X
368	Tri−State Hex Inverter	16										X
373	Tri−State Octal Transpare	20	X	X	X	X	X		X			
374	Tri−State Octal D Flip/Flop	20	X	X	X	X	X		X	X		X
375	Quad Latch	16										X
377	Octal D Type Flip/Flop with Enable	20			X	X	X					X
378	Hex D Flip/Flop with Enable	16										X
379	Hex D Flip/Flop with Enable, Inverted Outputs	16					X					
390	Dual Decade Ripple Counter	16										X
393	Dual 4 Bit Binary Ripple Counter	14	X									X

Type	Description	Pins	74 HC	74 HCT	74 AC	74 ACT	74 F	4000	74 ALS	74 AS	74	74 LS
399	Quad 2 Input Multiplexer	16			X	X	X					X
423	Multivibrator (no trigger from clear)	16										X
442	Quad tridirectional bus drivers	20										X
465	Octal Buffer Tri−State Output	20										X
518	Octal Comparator	20							X			
520	Octal Comparator	20			X	X	X		X			
521	Octal Comparator	20			X	X	X		X			
533	Octal Transparent latch, Tri−State Output	20				X	X		X			
534	Inverting Octal D Flip/Flop	20				X	X		X			
540	Octal Buffer and Line DR/RX Inverted	20	X	X	X		X		X			X
541	Octal Buffer and Line DR/RX Non−Inverting	20	X	X	X	X	X		X			X
543	Octal Registered Transceiver	24					X					
545	Octal Tri−State Transceiver	24					X					
561	Synchronous 4−bit counters, 3−state	20							X			
563	Inverted Octal D Type Latch	20				X			X			
564	Octal D Type Latch	20					X		X			
569	Synchronous Bi−Directional Binary Counter	20							X			
573	Octal D Type Latches	20	X	X	X	X	X		X	X		
574	Octal D Type Flip/Flop	20	X	X	X	X	X		X	X		
575	Octal D Type Flip/Flop	20							X			
576	Inverting Octal D Type Flip/Flop	20							X			
577	Octal D Type Flip/Flop	20							X			
579	8 Bit Up/Down Counter	20					X					
580	Inverting Octal D Type Latch	20							X			
589	8 Bit Shift Register with Input Latch	16	X									
590	8 Bit Binary Counter, O/P Register 3−state	16										X
592	8 Bit Binary Counter with Input Register	16										X
593	8 Bit Binary Counter, O/P Register TS I/O	16										X
594	8 Bit Binary Counter, O/P Register TS I/O	16	X									X
595	8 Bit Shift Register with Output Latch	16	X									X
596	8 Bit Shift Register with Output Latch	16										X

Type	Description	Pins	74 HC	74 HCT	74 AC	74 ACT	74 F	4000	74 ALS	74 AS	74	74 LS
597	8 Bit Shift Register with Input Latch	16										X
598	8 Bit Shift Register with Input Latch	16										X
604	16 to 8 Multiplexer (High Speed)	28										
620	Octal Bus Transceivers	20							X			X
621	Octal Bus Transceivers	20							X			X
623	Octal Bus Transceivers	20							X			X
624	Voltage Controlled Oscillator	14										X
628	Voltage Controlled Oscillator	14										X
629	Voltage Controlled Oscillator	16										X
638	Octal Bus Transceivers	20							X			
639	Octal Bus Transceivers	20							X			
640	Octal Bus Transceiver	20							X	X		X
641	Octal Bus Transceiver	20							X			X
642	Octal Bus Transceiver	20							X			X
645	Octal Bus Transceiver	20							X			X
646	Octal TS Bus Transceiver Non–Inverting	24			X	X	X		X	X		X
648	Octal TS Bus Transceiver Inverting	24			X		X		X	X		X
651	Octal Bus Transceiver Non–Inverting	24							X	X		
652	Octal Bus Transceiver Non–Inverting	24				X	X		X	X		X
653	Octal Bus Transceiver	24							X			
654	Octal Bus Transceiver	24							X			
666	Octal D Type Latch, 3–state	24							X			
667	Octal D Type Latch, 3–state	24							X			
669	Synchronous 4–Bit Up/Down Binary Counter	16										X
670	Tri–State 4 4 Register File	16										X
673	16 Bit Shift Register, 16 Bit Parallel Output	24					X					X
674	16 Bit Shift Register, 16 Bit Parallel Output	24										X
675	16 Bit Shift Register, 16 Bit Parallel Output	24					X					
676	16 Bit Shift Register, 16 Bit Parallel Output	24					X					
679	12–Bit Address Comparators	20							X			
682	8 Bit Magnitude Comparator	20										X

Type	Description	Pins	74HC	74HCT	74AC	74ACT	74F	4000	74ALS	74AS	74	74LS
684	8 Bit Magnitude Comparator	20										X
688	8 Bit Magnitude Comparator	20	X						X			X
697	Synchronous 4−Bit Up/Down Binary Counter	20										X
760	Octal Buffer &Line Driver, OC	20							X			
779	8−bit bidirectional binary counter, 3−state	16					X					
804	Hex 2−input NAND drivers	20							X			
805	Hex 2−input NOR drivers	20							X			
821	10 Bit D Type Flip/Flop	24			X	X	X					
823	9 Bit D Type Flip/Flop	24				X	X					
825	8 Bit D Type Flip/Flop	24				X	X					
827	10 Bit Buffer/Line Driver	24					X					
832	Hex 2−input OR drivers	20							X			
841	10 Bit Transparent Latch	24				X			X			
843	9 Bit Transparent Latch	24				X			X			
857	Hex 2−line to 1−line multiplexer, 3−state	24							X			
867	Synchronous 8−bit up/down Binary Counter	20							X			
869	Synchronous 8−bit up/down Binary Counter	20							X			
870	Dual 16 by 4 register files	24							X			
873	Dual 4−bit latches, 3−state	24							X			
874	Dual 4−bit D−type flip flops, 3−state	24							X			
876	Dual 4−bit D−type flip flops, 3−state	24							X			
899	9−Bit Latchable Transceiver with parity checker/generator	28			X	X						
902	Hex Buffer (TTL Interface)	14										
906	Open Drain Buffer (Active Pull Down)	14										
908	Dual High Voltage CMOS Driver	8										
912	Display Controller 6 Digit, 8 Segment	28										
914	Hex Schmitt Trigger, External Voltage Inputs	14										
922	16 Key Keyboard Encoder	18										
923	20 Key Keyboard Encoder	20										

Type	Description	Pins	74 HC	74 HCT	74 AC	74 ACT	74 F	4000	74 ALS	74 AS	74	74 LS
925	4 Digit Counter, Multiplexed 7 Segment Drive	16										
990	Octal D−type Latch, 3−state	20							X			
992	9−Bit D−type readback latch, 3−state	24							X			
994	10−Bit D−type readback latch	24							X			
996	Octal D−type readback latch	24							X			

Type	Description	Pins	74 HC	74 HCT	74 AC	74 ACT	74 F	4000	74 ALS	74 AS	74	74 LS
4000	Dual 3 Input NOR Gate plus Inverter	14										
4001	Quad 2 Input NOR Gate	14						X				
4002	Dual 4 Input NOR Gate	14										
4006	18 Bit Static Shift Register	14										
4007	Dual Complementary Pair plus Inverter	14										
4008	4 Bit Full Adder	16										
4009	Hex Inverting Buffers	16										
4010	Hex Buffer (Non − Inverting)	16										
4011	Quad 2 Input NAND Gate	14						X				
4012	Dual 4 Input NAND Gate	14										
4013	Dual D Flip/Flop with Set/Reset	14						X				
4014	8 Bit Static Shift Register	16										
4015	Dual 4 Bit Static Shift Register	16										
4016	Quad Bi − Lateral Switch	14						X				
4017	Decade Counter/Divider	16										
4018	Presettable Divide − by − N Counter	16										
4019	Quad AND/OR Select Gate	16										
4020	14 Stage Ripple Carry Binary Counter	16	X									
4021	8 Bit Static Shift Register	16						X				
4022	Divide − by − 8 Counter/Divider	16										
4023	Triple 3 Input NAND Gate	14										
4024	7 Bit Binary Counter	14										
4025	Triple 3 Input NOR Gate	14										
4026	Bcd Decade Counter, 7 − Segment Decoder	16										
4027	Dual J − K Flip/Flop	16										
4028	BCD − to − Decimal Decoder	16										
4029	Presettable Up/Down Binary/Decade Counter	16										
4030	Quad EX − OR Gate	14										
4031	64 − Stage Shipt Register	16										
4032	Triple Positive Logic Serial Adders	16										
4033	Bcd Decade Counter, 7 − Segment Decoder	16										

Type	Description	Pins	74 HC	74 HCT	74 AC	74 ACT	74 F	4000	74 ALS	74 AS	74	74 LS
4034	8−Bit Bi−Directional Shift Register	24										
4035	4 Bit Shift Register	16										
4038	Triple Negative Logic Serial Adders	16										
4040	12 Bit Binary Ripple Counter/Divider	16	X									
4041	Quad True/Complement Buffer	14										
4042	Quad D Latch	16										
4043	Quad Tri−State NOR R/S Latch	16										
4044	Quad Tri−State NAND R/S Latch	16										
4045	21−Stage Binary Counter/Divide W/Oscillator	16										
4046	Phase Locked Loop	16	X									
4047	Monostable/Astable Multivibrator	14						X				
4048	8−Input Expandable Multifuntion Gate	16										
4049	Hex Inverting Buffer	16	X					X				
4050	Hex Buffer	16	X					X				
4051	Single 8 Channel Multiplexer	16	X					X				
4052	Differential 4 Channel Multiplexer	16	X					X				
4053	Triple 2 Channel Multiplexer	16	X					X				
4054	BCD to 7 Segment LCD Decoder/Drive	16										
4055	Bcd To 7−Segment Decoder Lcd	16										
4056	Bcd To 7−Segment Decoder Lcd Driver	16										
4059	Programmable Divide−by−N Counter	24										
4060	14 Stage Ripple Carry Binary Counter	16	X					X				
4063	4 Bit Magnitude Comparator	16										
4066	Quad Bi−Lateral Switch	14	X					X				
4067	16 Channel Analogue Multiplexer/Demultiplexer	24										
4068	8 Input NAND Gate	14										
4069	Hex Inverter	14						X				
4070	Quad Exclusive OR Gate	14										
4071	Quad 2 Input OR Gate	14										
4072	Dual 4 Input OR Gate	14										

Type	Description	Pins	74 HC	74 HCT	74 AC	74 ACT	74 F	4000	74 ALS	74 AS	74	74 LS
4073	Triple 3 Input AND Gate	14										
4075	Triple 3 Input OR Gate	14										
4076	Tri−State Quad Latch	16										
4077	Quad Exclusive NOR Gate	14										
4078	8 Input NOR/OR Gate	14										
4081	Quad 2 Input AND Gate	14										
4082	Dual 4 Input AND Gate	14										
4085	Dual 2 Wide Input AND/OR Invert Gate	14										
4086	Expandable 4 Wide, 2 Input AND−0R−INVERT Gate	14										
4089	Cascadable 4−Bit Binary Rate Multiplier	16										
4093	Quad 2 Input NAND Schmitt Trigger	14					X					
4094	8 Stage Shift and Store Bus Register	16										
4095	Gated J−K Master Slave Flip−Flop	14										
4096	Gated J−K Master Slave Flip−Flop	14										
4097	Differential 8−Channel Mux/Demux	24										
4098	Retriggerable Dual Monostable Multivibrator	16										
4099	8 Bit Addressable Latch	16										
4104	Tri−State Quad Low Voltage to High Voltage Translator	16										
4106	Hex Schmitt Trigger (40106)	14										
4161	4 Bit Synchronous Progammable Binary Counter	16										
4163	4 Bit Synchronous Progammable Binary Counter	16										
4174	Hex D−type Flip−Flops	16										
4175	4 D−type Flip−Flops	16										
4194	4 Bit Bidirectional Shift Register	16										
4500	Industrial Control Unit	16										
4501	Dual 4−Input NAND, 2−Input NOR/OR gate	16										
4502	Strobed Hex Inverter/Buffer	16										

Type	Description	Pins	74 HC	74 HCT	74 AC	74 ACT	74 F	4000	74 ALS	74 AS	74	74 LS
4503	Tri-State Hex Buffer	16										
4504	Hex Level Shifter	16										
4506	Dual Expandable AND/OR Gate	16										
4508	Dual 4 Bit Latch	24										
4510	BCD Up/Down Counter	16										
4511	BCD to 7 Segment Decoder/Driver	16										
4512	8 Channel Data Selector	16										
4513	BCD-to-7-Segment Latch/Decoder/Driver with Ripple Blanking	18										
4514	4 Bit Latch/4 to 16 Line Decoder (High)	24	X									
4515	4 Bit Latch/4 to 16 Line Decoder (Low)	24										
4516	Binary Up/Down Counter	16										
4517	Dual 64 Bit Static Shift Register	16										
4518	Dual BCD Up Counter	16										
4519	4 Bit AND/OR Selector	16										
4520	Dual Binary Up Counter	16										
4521	24 Stage Frequency Divider	16										
4522	Divide-by-N Counter (BCD)	16										
4526	Divide-by-N Counter (Binary)	16										
4527	BCD Rate/Multiplier	16										
4528	Dual Retriggerable Resettable/Monostable Multivibrator	16										
4529	Dual 4-Channel Analog Data Selector	16										
4530	Dual 5-Input Majority Logic Gate	16										
4531	12-Bit Parity Tree	16										
4532	8 Input Priority Encoder	16										
4534	Real Time 5-Decade Counter	24										
4536	Programmable Timer	16										
4538	Dual Monostable Multivibrator	16	X					X				
4539	Dual 4 Channel Data Selector/Multiplexer	16										

Type	Description	Pins	74 HC	74 HCT	74 AC	74 ACT	74 F	4000	74 ALS	74 AS	74	74 LS
4541	Programmable Oscillator Timer	14						X				
4543	BCD to 7 Segment Latch/Decoder/Driver for LCDs	16										
4544	BCD−to−7−Segment Latch/Decoder/Driver with Ripple Blanking	18										
4547	High Current BCD−to−7−Segment Decoder/Driver	16										
4551	Quad 2 Input Analogue Multiplexer	16										
4553	3 Digit BCD Counter	16										
4555	Dual Binary−to−1 of 4 Decoder/Demultiplexer	16										
4556	Dual Binary−to−1 of 4 Decoder/Demultiplexer (Inverting)	16										
4557	1 −to−64 Bit Variable Length Shift Register											
4558	BCD−to−7−Segment Decoder	16										
4559	Successive Approximation Register	16										
4560	NBCD Adder	16										
4561	9's Complementer	14										
4562	128−Bit Static Shift Register	14										
4568	Phase Comparator and Programmable Counter	16										
4569	Programmable Dual Binary/BCD Counter	16										
4572	Hex Gate	16										
4580	4×4 Multiport Register	24										
4582	Look−Ahead Carry Generator	16										
4583	Dual Schmitt Trigger	16										
4584	Hex Schmitt Trigger	14										
4585	4 Bit Magnitude Comparatof	16										
4598	8 Bit Bus Compatible Addressable Latch	18										
4599	8 Bit Addressable Latch	18										
4724	8−Bit Addressable Latch	16										
40085	Cascadable 4−Bit Magnitude Comparator	16										
40097	Hex 3 State Buffer	16										

Type	Description	Pins	74 HC	74 HCT	74 AC	74 ACT	74 F	4000	74 ALS	74 AS	74	74 LS
40098	Hex 3 State Inverting Buffer	16										
40100	32 – Stage Static Left/Right Shift Register	16										
40101	9 – Bit Parity Generator/Checker	14										
40102	Presettable Sync. Bcd 2 – Decade Down Conter	16										
40103	Presettable 8 Bit Binary Down Counter	16										
40104	4 – Bit Bidir. Univ. Shift Register	16										
40105	16 – Word By 4 – Bit Fifo Register	16										
40106	Hex Schmitt Trigger	14						X				
40107	Dual 2 Input NAND Gate (Driver)	8										
40108	4×4 Multiport Register	24										
40109	Quad Low – To – High Voltage Level Shifters	16										
40110	Bcd Decade Up/Down Counter/Decoder	16										
40114	64 Bit (16×4)RAM	16										
40147	10 Line to 4 Line DCD Priority Encoder	16										
40160	Sync. Bcd Decade Counter W/Clear	16										
40161	Synchronous Binary Counter	18										
40163	Sync. 4 – Bit Binary Counter W/Sync. Clear	16										
40164	Hex D – Type Flip Flops Single Rail Output	16										
40174	Hex D Flip/Flop	16										
40175	Quad D Flip/Flop	16										
40181	Arithmetic Logic Unit/Function Generator	24										
40182	Look Ahead Carry Generator	16										
40192	Decade Up/Down Counter	16										
40193	Binary Up/Down Counter	16										
40194	4 – Bit Bidir. Univ. Shift Register	16										
40244	Octal Buffers with Tri – State Outputs	20										
40245	Octal Bus Transceiver with Tri – State Output	20										
40257	Quad 2 – To – 1 Line Data Selector/Mux	16										
40373	Octal Transparent Latch with Tri – State Outputs	20										
40374	Octal D Type Flip/Flop with Tri – State Outputs	20										

박준식 ─────────────────────────────────

▌약력

동아대학교 전자공학과 졸업
경남대학교 대학원 전자공학과 졸업
강원대학교 대학원 전자공학과 졸업
한국통신(KT) 근무
KT 인포텍(KTI) 근무
한국폴리텍6 구미대학 정보통신시스템과 근무
현재: 강릉영동대학 의료전자과 근무

▌주요저서

『데이터통신총론』(1997. 3, 도서출판 기다리)
『디지털공학 기초실험』(1998. 2, 기한재)
『엑셀과 파워포인트』(2001. 3, 웅보출판사)
『실용컴퓨터』(2001. 3, 웅보출판사)
『알기 쉽게 배우는 엑셀과 파워포인트』(2007. 1, 청송출판사)
『사이버포렌식 총론』(2006. 11, 청송출판사)

초보자도 쉽게 이해할 수 있는

디지털회로 기초실험

초판인쇄 | 2009년 2월 16일
초판발행 | 2009년 2월 16일

지은이 | 박준식
펴낸이 | 채종준
펴낸곳 | 한국학술정보㈜
주 소 | 경기도 파주시 교하읍 문발리 513-5 파주출판문화정보산업단지
전 화 | 031) 908-3181(대표)
팩 스 | 031) 908-3189
홈페이지 | http://www.kstudy.com
E-mail | 출판사업부 publish@kstudy.com

등 록 |
가 격 | 35,000원

ISBN 978-89-534-1259-0 93560 (Paper Book)
      978-89-534-1260-6 98560 (e-Book)